Advanced Leadership in Qualified Precepting

For Respiratory Care Practitioners

Author Amber Thurman RCP, MPH, MBA

Published By Verbiage Basics

TABLE OF CONTENTS

Overview

This comprehensive material is designed for experienced Respiratory Care Practitioners seeking to enhance their leadership skills in the field, with a specific focus on qualified precepting of respiratory care students. As an integral part of healthcare teams, respiratory care practitioners often take on the role of preceptors, guiding and mentoring students in their clinical rotations.

The material delves into advanced concepts of leadership, emphasizing the importance of qualified precepting in shaping the next generation of respiratory care professionals. Participants will explore strategies for effective communication, mentorship, and fostering a positive learning environment. Key topics include assessing student competencies, providing constructive feedback, and creating personalized learning plans to support individual student needs.

Through a combination of theoretical discussions, case studies, and practical simulations, participants will develop the skills necessary to excel as preceptors in respiratory care settings. The material will also address ethical considerations, cultural competence, and the integration of evidence-based practices in precepting.

By the end of the book, participants will be equipped with the knowledge and tools to confidently precept respiratory care students, ensuring a seamless transition from the study to real-world clinical practice. This advanced leadership material is tailored to empower respiratory care practitioners to play a pivotal role in the education and development of future healthcare professionals.

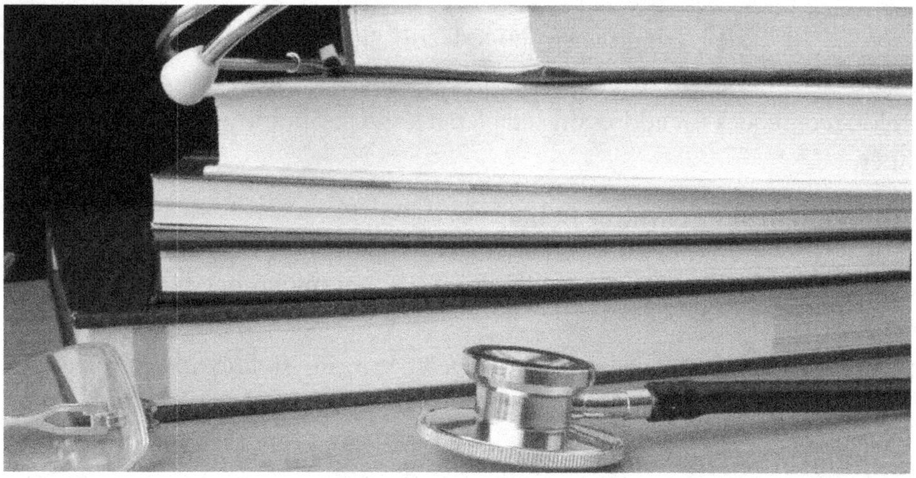

KEY LEARNING OBJECTIVES

1. **Understand Advanced Leadership Concepts:**

   — Explore advanced theories and principles of leadership relevant to respiratory care practitioners in precepting roles.
   — Analyze leadership styles and their application to effective precepting in respiratory care settings.

2. **Enhance Communication and Mentorship Skills:**

   — Develop advanced communication techniques for fostering positive relationships with respiratory care students.
   — Hone mentorship skills to guide students through their clinical experiences and professional development.

3. **Assess and Evaluate Student Competencies:**

   — Learn comprehensive strategies for assessing and evaluating the clinical competencies of respiratory care students.
   — Utilize objective measures to identify areas of improvement and success in student performance.

4. **Provide Constructive Feedback:**

   — Master the art of delivering constructive feedback to students, fostering a growth mindset and continuous improvement.
   — Develop techniques for addressing challenges and celebrating achievements in a supportive manner.

5. **Create Individualized Learning Plans:**

   — Design personalized learning plans tailored to the unique needs and goals of respiratory care students.
   — Implement strategies to accommodate diverse learning styles and optimize the educational experience.

6. **Explore Ethical Considerations in Precepting:**

— Examine ethical dilemmas that may arise in the precepting relationship and develop strategies for ethical decision-making.
— Foster a culture of professionalism, integrity, and respect within the precepting dynamic.

7. **Promote Cultural Competence in Precepting:**

— Understand the importance of cultural competence in respiratory care education.
— Develop skills to create an inclusive and culturally sensitive learning environment for students from diverse backgrounds.

8. **Integrate Evidence-Based Practices in Precepting:**

— Explore evidence-based practices in respiratory care education and their application to precepting.
— Incorporate research findings into the development of effective precepting strategies.

9. **Simulate Real-World Precepting Scenarios:**

— Engage in practical simulations to apply learned concepts in realistic precepting scenarios.
— Receive constructive feedback on precepting skills through simulation-based exercises.

10. **Empower Practitioners for Future Precepting Roles:**

— Equip participants with the knowledge and confidence to excel as preceptors in respiratory care settings.
— Foster a sense of responsibility and leadership in shaping the next generation of respiratory care professionals.

GOALS

1. **Leadership Mastery:**
    - Develop an advanced understanding of leadership theories and principles specific to respiratory care practitioners in precepting roles.
2. **Effective Communication and Mentorship:**
    - Enhance communication and mentorship skills to establish positive relationships with respiratory care students and foster their professional growth.
3. **Competency Assessment Expertise:**
    - Acquire comprehensive strategies for assessing and evaluating the clinical competencies of respiratory care students.
4. **Constructive Feedback Proficiency:**
    - Master the delivery of constructive feedback to guide students toward continuous improvement and success.
5. **Individualized Learning Plans Creation:**
    - Design personalized learning plans tailored to the unique needs and goals of respiratory care students.
6. **Ethical Precepting Practices:**
    - Examine and navigate ethical considerations in the precepting relationship, fostering a culture of professionalism and integrity.
7. **Cultural Competence Integration:**
    - Understand and apply cultural competence principles in respiratory care education, creating an inclusive learning environment.
8. **Evidence-Based Precepting:**
    - Integrate evidence-based practices into the precepting process, staying informed about the latest research in respiratory care education.
9. **Simulation-Based Skill Application:**

- Apply learned concepts in realistic precepting scenarios through practical simulations, receiving feedback to enhance skills.

10. **Confident and Empowered Preceptors:**

    - Equip participants with the knowledge, skills, and confidence to excel as preceptors in respiratory care settings, contributing to the development of future professionals.

CHAPTER 1: LEADERSHIP FOUNDATIONS IN RESPIRATORY CARE

Objective: 1

- Introduction to leadership theories and principles
- Application of leadership styles in respiratory care settings
- The role of advanced leadership in precepting respiratory care students

Introduction to Leadership Theories

Overview: Introducing fundamental leadership theories applicable to respiratory care practitioners engaged in precepting roles. Understanding these theories is essential for effective leadership in guiding and mentoring respiratory care students.

Content:

1. *Trait Theories:* Examining the traits associated with effective leadership in healthcare settings.
2. *Behavioral Theories:* Understanding how specific behaviors contribute to successful leadership in clinical environments.
3. *Contingency Theories*: Exploring the situational aspects influencing leadership effectiveness.
4. *Transformational Leadership:* Introducing the concept of transformational leadership and its impact on team dynamics.

Trait Theories

Trait theories of leadership focus on identifying and analyzing the specific traits or characteristics that distinguish effective leaders. In the context of respiratory care, practitioners with traits such as empathy, adaptability, and effective communication skills may excel in leadership roles. Trait theories help in understanding the personal qualities associated with successful precepting, guiding respiratory care students through their clinical experiences.

Behavioral Theories

Behavioral theories emphasize the actions and behaviors of leaders rather than inherent traits. In respiratory care, effective leaders may exhibit behaviors such as clear communication, active listening, and the ability to provide constructive feedback. Behavioral theories contribute to shaping precepting styles and strategies that positively influence the learning experiences of respiratory care students.

Contingency Theories

Contingency theories suggest that the effectiveness of leadership styles depends on the situational context. In respiratory care, preceptors may need to adapt their leadership approach based on the unique needs of individual students or specific clinical scenarios. Understanding contingency theories allows respiratory care practitioners to tailor their precepting strategies to different situations for optimal student guidance.

Transformational Leadership

Transformational leadership involves inspiring and motivating others to achieve higher levels of performance. In respiratory care, a transformational leader may foster a positive learning environment, encourage innovation, and instill a sense of purpose in students. Transformational leadership is particularly relevant in precepting, where guiding and shaping the future of respiratory care professionals requires inspirational and forward-thinking leadership approaches.

Application of Leadership Styles in Respiratory Care

Overview: Focusing on the practical application of leadership styles within the context of respiratory care settings. Participants will explore how assorted styles can be leveraged to enhance precepting effectiveness.

Content:

1. *Autocratic Leadership:* Examining the use of authoritative direction in specific clinical scenarios.
2. *Democratic Leadership:* Understanding collaborative decision-making and its role in respiratory care precepting.
3. *Laissez-Faire Leadership:* Exploring the hands-off approach and its appropriateness in certain situations.
4. *Situational Leadership:* Applying leadership styles based on the unique needs of respiratory care students.

Autocratic Leadership

Autocratic leadership is characterized by a leader making decisions independently with little input from others. In respiratory care, an autocratic approach may be necessary in urgent or critical situations where quick and decisive action is required. For example, during a respiratory emergency, a respiratory care practitioner in a leadership role may need to take charge and make immediate decisions to ensure patient safety.

Democratic Leadership

Democratic leadership involves collaboration and decision-making through group consensus. In respiratory care, democratic leadership may be employed when planning and implementing changes in clinical protocols or discussing strategies for optimizing patient care. This approach encourages input from respiratory care team members, fostering a sense of shared responsibility and inclusivity in decision-making processes.

Laissez-Faire Leadership

Laissez-Faire leadership is characterized by a hands-off approach, allowing team members considerable freedom in decision-making. In respiratory care, a laissez-faire leadership style might be effective when working with experienced and self-motivated respiratory care practitioners. This approach encourages autonomy and self-directed learning, which can be beneficial in certain educational and clinical scenarios.

Situational Leadership

Situational leadership involves adapting leadership styles based on the specific needs of the situation or individuals involved. In respiratory care, a situational leader may adjust their approach depending on the complexity of a clinical case or the developmental stage of a respiratory care student. This flexibility allows the leader to tailor their guidance and support to best meet the demands of the respiratory care environment.

The Role of Advanced Leadership in Precepting

Overview: By delving into the significance of advanced leadership in the precepting dynamic. It highlights the crucial role respiratory care practitioners play in shaping the next generation of professionals through mentorship and guidance.

Content:

1. *Leadership as a Guiding Force:* Understanding how advanced leadership positively impacts the precepting relationship.
2. *Mentorship and Leadership:* Exploring how mentorship aligns with advanced leadership principles in respiratory care.
3. *Leadership Strategies for Precepting Success:* Identifying specific leadership strategies to optimize the precepting experience for respiratory care students.

Leadership as a Guiding Force

Leadership as a guiding force in respiratory care involves providing direction, support, and inspiration to shape the development of respiratory care professionals. In this context, leaders play a pivotal role in guiding the trajectory of their team members and respiratory care students. The guiding force of leadership ensures that practitioners are mentored effectively, fostering a positive and constructive learning environment.

Mentorship and Leadership

Mentorship and leadership in respiratory care go hand in hand, emphasizing the importance of experienced practitioners guiding and supporting the growth of students. Mentorship involves sharing knowledge, providing feedback, and serving as a role model for aspiring respiratory care professionals. Effective mentorship within a leadership framework enhances the educational experience, contributing to the overall development of competent and confident practitioners.

Leadership Strategies for Precepting Success

Leadership strategies for precepting success involve employing intentional approaches to guide respiratory care students effectively. These strategies encompass a range of skills, including communication, feedback delivery, and the ability to create a supportive learning environment. Successful precepting leaders adapt their strategies based on the unique needs of individual students and the dynamic nature of respiratory care settings, ensuring a positive and impactful precepting experience.

Chapter 2: Communication and Mentorship Skills

Objectives: 2

- Effective communication strategies for precepting
- Developing mentorship relationships with respiratory care students
- Addressing challenges in communication and mentorship

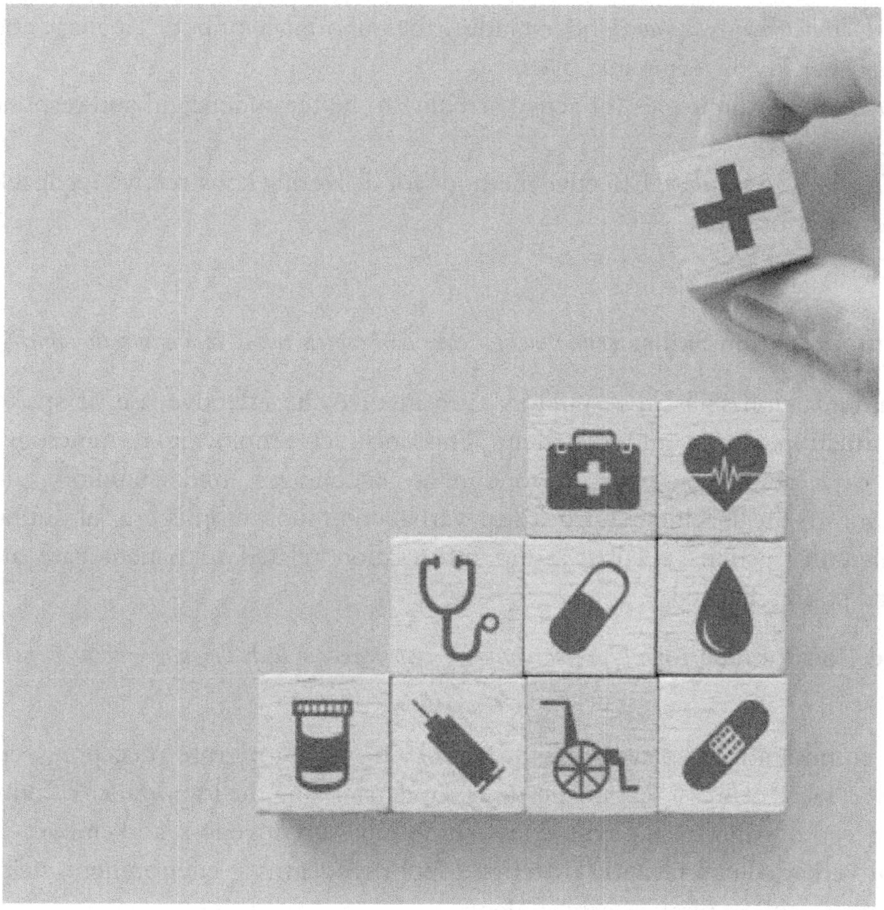

Communication Strategies for Precepting

Overview: Now focusing on developing effective communication strategies crucial for successful precepting in respiratory care. Clear and impactful communication is essential for building positive relationships with students and facilitating their learning experiences.

Content:

1. *Verbal Communication Skills:* Techniques for clear and concise verbal communication during precepting.
2. *Non-Verbal Communication*: Understanding the importance of body language and other non-verbal cues in conveying messages.
3. *Active Listening:* Strategies for active listening to better understand and respond to student needs.
4. *Feedback Communication:* Effective methods for delivering constructive feedback to enhance student performance.

Verbal Communication Skills: *Techniques for clear and concise verbal communication during precepting*

Verbal communication skills in respiratory care involve the effective use of spoken words to convey information clearly and concisely. This includes employing terminology familiar to respiratory care students, using a tone that encourages understanding, and adapting communication to the specific context. Clear verbal communication is crucial during precepting to ensure that instructions, feedback, and information related to patient care are accurately conveyed and comprehended.

Non-Verbal Communication: *Understanding the importance of body language and other non-verbal cues in conveying messages*

Non-verbal communication plays a significant role in respiratory care precepting, encompassing body language, facial expressions, and gestures. Understanding the importance of non-verbal cues is vital for preceptors as these cues can convey emotions, attentiveness, and empathy. Maintaining positive non-verbal communication fosters a supportive learning environment, helping to build trust and rapport with respiratory care students.

Active Listening: *Strategies for active listening to better understand and respond to student needs*

Active listening involves fully engaging with the speaker, demonstrating attentiveness, and providing feedback. In respiratory care precepting, active listening is crucial for understanding the concerns, questions, and perspectives of students. Preceptors can employ strategies such as paraphrasing, asking clarifying questions, and demonstrating empathy to enhance active listening.

This skill contributes to effective communication, promotes a collaborative atmosphere, and ensures that student needs are acknowledged and addressed.

Feedback Communication

Feedback communication in respiratory care precepting involves providing constructive feedback to students regarding their performance, clinical skills, and decision-making. This includes offering specific praise for strengths and providing guidance on areas for improvement. Effective feedback communication is a two-way process, encouraging students to reflect on their performance and fostering a culture of continuous improvement. Constructive feedback is essential for shaping the development of respiratory care students and preparing them for real-world clinical scenarios.

Mentorship Development in Respiratory Care

Overview: Exploring the significance of mentorship in respiratory care and provides insights into developing mentorship skills. Effective mentorship is crucial for guiding students through their educational journey and fostering their professional growth.

Content:

1. *Role of the Mentor:* Defining the responsibilities and expectations of a mentor in respiratory care.
2. *Building Trust and Rapport:* Strategies for establishing trustful and supportive relationships with mentees.
3. *Professional Development:* Mentoring for career advancement and ongoing professional growth in respiratory care.
4. *Mentorship Models:* Exploring different mentorship models and their applicability in healthcare settings.

Role of the Mentor: *Defining the responsibilities and expectations of a mentor in respiratory care*

The role of the mentor in respiratory care involves guiding, advising, and supporting individuals in their professional development. Mentors in respiratory care have the responsibility of sharing their knowledge and expertise with mentees, helping them navigate their educational and clinical experiences. Responsibilities may include providing career guidance, offering insights into the field, and serving as a role model. Clear expectations and communication between mentors and mentees are essential for a successful mentorship relationship.

Building Trust and Rapport: Strategies for establishing trustful and supportive relationships with mentees

Building trust and rapport is foundational to effective mentorship in respiratory care. Trustful relationships create a supportive environment where mentees feel comfortable seeking guidance and sharing challenges. Strategies for building trust include active listening, open communication, and demonstrating empathy. Establishing rapport involves creating a positive and collaborative atmosphere, fostering a mentorship dynamic that encourages mentees to engage in meaningful professional development.

Professional Development: *Mentoring for career advancement and ongoing professional growth in respiratory care*

Mentoring in respiratory care extends beyond immediate educational needs, encompassing long-term professional development. Mentors play a key role in guiding mentees toward career advancement, offering insights into potential career paths, and providing advice on skill development. Professional development in mentorship involves setting goals, creating action plans, and supporting mentees in their journey toward becoming competent and confident respiratory care professionals.

Mentorship Models: *Exploring different mentorship models and their applicability in healthcare settings*

Various mentorship models exist, and their applicability in healthcare settings, including respiratory care, depends on the goals and preferences of the mentor-mentee pair. Traditional one-on-one mentorship, group mentorship, and peer mentoring are examples of mentorship models. Exploring these models allows mentors and mentees to choose an approach that aligns with their needs and preferences, enhancing the effectiveness and versatility of the mentorship relationship in the respiratory care context.

Discussion: *Addressing Communication Challenges in Precepting*

Overview: This discussion session aims to address common communication challenges encountered in the precepting relationship. Participants will share experiences, explore solutions, and engage in collaborative problem-solving to enhance communication effectiveness.

Discussion Points:

1. *Identifying Communication Barriers:* Recognizing common obstacles to effective communication in precepting.
2. *Case Studies:* Analyzing real-life scenarios to understand and address communication challenges.
3. *Interactive Strategies:* Engaging participants in role-playing and simulation exercises to practice overcoming communication obstacles.

Discussion Points: *Identifying Communication Barriers: Recognizing common obstacles to effective communication in precepting*

Discussion points related to identifying communication barriers in respiratory care precepting involve recognizing and understanding obstacles that may hinder effective communication. Participants discuss familiar challenges such as misinterpretation of instructions, language barriers, or differing communication styles. Identifying these barriers is crucial for developing strategies to overcome them, ensuring that communication remains clear and conducive to successful precepting.

Case Studies: *Analyzing real-life scenarios to understand and address communication challenges*

The inclusion of case studies in discussions allows participants to analyze real-life scenarios where communication challenges in respiratory care precepting may have occurred. By examining these cases, participants gain insights into the complexities of communication, potential pitfalls, and effective strategies for resolution. Case studies provide a practical and contextual understanding of communication challenges, enabling participants to apply critical thinking skills in their precepting roles.

Interactive Strategies: *Engaging participants in role-playing and simulation exercises to practice overcoming communication obstacles*

Interactive strategies, such as role-playing and simulation exercises, offer participants hands-on experience in addressing communication obstacles. Engaging in simulated scenarios relevant to respiratory care precepting allows participants to practice effective communication techniques, receive immediate feedback, and refine their skills. These interactive sessions contribute to the development of practical communication strategies that can be applied in real-world precepting situations.

In the context of respiratory care, these discussion points and interactive strategies are vital for preceptors to enhance their communication skills, navigate challenges effectively, and create a positive and conducive learning environment for students. Effective communication is essential in ensuring that critical information is conveyed accurately, fostering a collaborative and supportive atmosphere in respiratory care education.

Chapter 3: Competency Assessment Strategies

Objectives: 3

- Methods for assessing clinical competencies in respiratory care students
- Objective measures and tools for competency evaluation
- Continuous assessment for ongoing student development

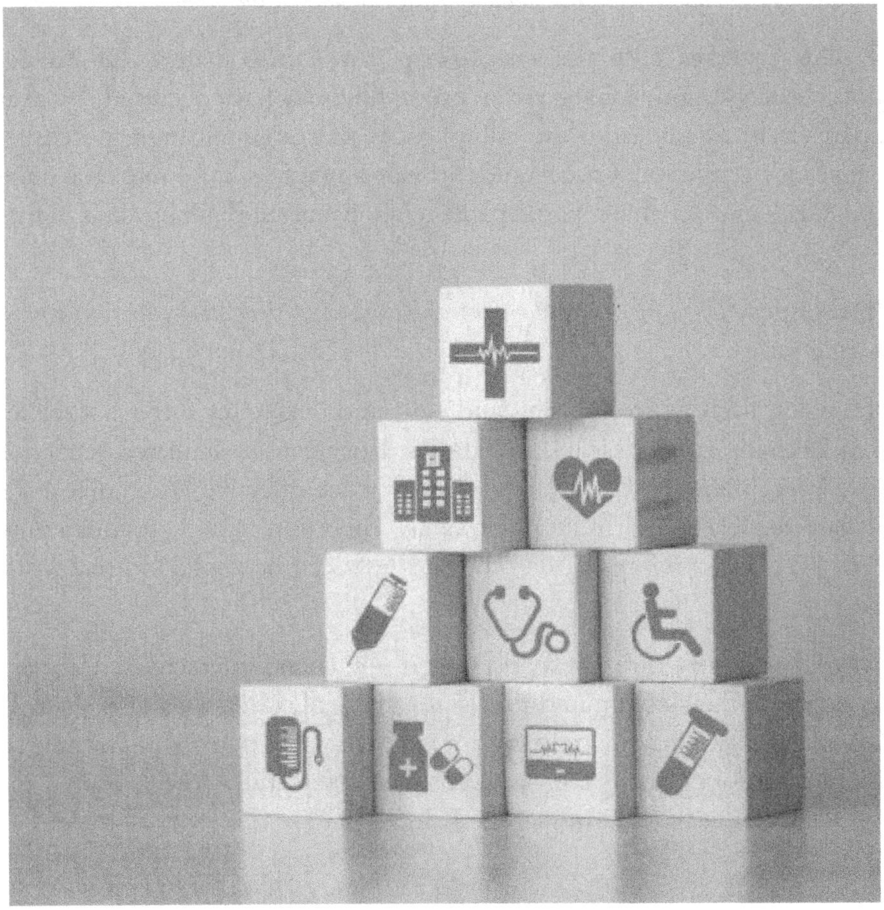

Methods for Assessing Clinical Competencies

Overview: Focus on exploring various methods for assessing clinical competencies in respiratory care. Assessing competencies is crucial for ensuring that students develop the necessary skills and knowledge to excel in their roles as respiratory care practitioners.

Content:

1. *Observational Assessment:* Using direct observation to evaluate students' clinical performance.
2. *Performance-Based Assessments:* Designing practical assessments to measure specific competencies.
3. *Simulated Scenarios:* Incorporating realistic scenarios in simulated environments to assess competencies.
4. Checklists and Rubrics: Utilizing structured checklists and rubrics for systematic competency evaluation.

Observational Assessment: *Using direct observation to evaluate students' clinical performance*

Observational assessment in respiratory care involves the direct observation of students' clinical performance in real-world settings. This method allows preceptors to assess students' practical skills, decision-making abilities, and overall competence during patient care. Observational assessment is valuable for providing immediate feedback, identifying areas for improvement, and ensuring that students can effectively apply theoretical knowledge to clinical practice in respiratory care settings.

Performance-Based Assessments: *Designing practical assessments to measure specific competencies*

Performance-based assessments in respiratory care are designed to measure specific competencies through practical tasks and scenarios. These assessments may include tasks such as administering respiratory treatments, conducting patient assessments, or responding to emergency situations. By focusing on hands-on skills and critical thinking, performance-based assessments ensure that students can apply theoretical knowledge to real-world situations, demonstrating their readiness for professional practice in respiratory care.

Simulated Scenarios: *Incorporating realistic scenarios in simulated environments to assess competencies*

Simulated scenarios in respiratory care involve creating lifelike situations in controlled environments to assess students' competencies. These scenarios may range from routine patient care to emergency situations. Simulated scenarios allow students to apply theoretical knowledge in a safe and supervised setting, providing preceptors with insights into their clinical decision-making, communication skills, and ability to handle various respiratory care challenges.

Checklists and Rubrics: *Utilizing structured checklists and rubrics for systematic competency evaluation*

Checklists and rubrics are structured tools used in respiratory care to systematically evaluate students' competencies. Checklists outline specific tasks or skills, allowing preceptors to mark off completed items during observation. Rubrics provide a more detailed evaluation, assigning scores or ratings to different aspects of performance. These tools offer objectivity in competency evaluation, ensuring that assessments are consistent and aligned with established standards in respiratory care education.

Objective Measures and Tools for Competency Evaluation

Overview: Examine objective measures and tools that can be employed for competency evaluation in respiratory care. Objective assessment tools provide quantifiable data and enhance the accuracy and reliability of competency evaluations.

Content:

1. *Standardized Testing:* Implementing standardized tests to assess knowledge-based competencies.
2. *Skills Checklists:* Utilizing structured checklists to evaluate specific skills and procedures.
3. *Objective Structured Clinical Examination (OSCE):* Incorporating OSCEs for comprehensive clinical competency evaluation.
4. *Electronic Health Records (EHR) Assessment:* Using EHR data to evaluate competency in documentation and information management.

Standardized Testing: *Implementing standardized tests to assess knowledge-based competencies*

Standardized testing in respiratory care involves the use of consistent and uniform assessments to evaluate knowledge-based competencies. These tests typically cover theoretical aspects of respiratory care, such as anatomy, physiology, and medical principles. Standardized testing ensures that all students are evaluated using the same criteria, allowing for fair and objective assessment of their knowledge, and understanding in respiratory care.

Skills Checklists: *Utilizing structured checklists to evaluate specific skills and procedures*

Skills checklists in respiratory care are structured tools that preceptors use to evaluate specific skills and procedures performed by students. These checklists outline step-by-step tasks, allowing preceptors to systematically assess students' proficiency in hands-on clinical skills. Skills checklists are valuable for providing targeted feedback and ensuring that students can competently perform essential procedures in respiratory care settings.

Objective Structured Clinical Examination (OSCE): *Incorporating OSCEs for comprehensive clinical competency evaluation*

Objective Structured Clinical Examination (OSCE) is a method of evaluating comprehensive clinical competencies in respiratory care. OSCEs involve multiple stations where students rotate through various scenarios, each designed to assess different aspects of clinical practice. Preceptors use standardized checklists or rubrics to evaluate students' performance, communication skills, and decision-making in realistic clinical situations. OSCEs provide a holistic approach to competency evaluation in respiratory care.

Electronic Health Records (EHR) Assessment: Using EHR data to evaluate competency in documentation and information management

Electronic Health Records (EHR) assessment in respiratory care involves evaluating students' competency in using digital platforms for documentation and information management. This assessment method focuses on the ability to accurately record patient information, manage electronic records, and navigate EHR systems efficiently. Proficiency in EHR usage is crucial for respiratory care practitioners to ensure accurate and organized patient information, making this assessment relevant to contemporary healthcare practices.

Practical Exercise: Continuous Competency Assessment

Overview: Practical exercise involves hands-on application of continuous competency assessment techniques in respiratory care. Participants engage in scenarios, utilizing various assessment methods and tools discussed in previous lectures.

Exercise Content:

1. *Scenario-Based Assessment*: Participants navigate simulated scenarios to assess competencies.
2. *Utilization of Objective Measures*: Participants apply objective measures and tools in competency evaluations.
3. *Feedback and Reflection*: Discussion and reflection on the effectiveness of continuous competency assessment strategies.

Exercise Content for a Scenario-Based Assessment: *Participants navigate simulated scenarios to assess competencies*

Overview: In this scenario-based assessment, participants will engage in simulated respiratory care scenarios to evaluate their competencies in handling real-world clinical situations. The exercise aims to assess participants' ability to apply theoretical knowledge, make critical decisions, and demonstrate effective clinical skills in simulated environments.

Scenarios:

1. **Acute Respiratory Distress Syndrome (ARDS) Management:** Participants will navigate a simulated scenario involving a patient experiencing ARDS. They will be assessed on their ability to assess the situation, implement appropriate interventions, and communicate effectively with the patient and healthcare team.
2. **Emergency Ventilation Troubleshooting:** Participants will encounter a simulated scenario where a mechanical ventilation system malfunctions. The assessment will focus on their capacity to troubleshoot the issue, implement corrective measures, and ensure patient safety during ventilation support.
3. **Patient Education and Inhaler Technique:** A scenario will be presented involving a patient with a chronic respiratory condition. Participants will be evaluated on their ability to educate the patient about their condition, demonstrate proper inhaler technique, and assess the patient's understanding.

Assessment Criteria:

- Clinical decision-making
- Effective communication with patients and healthcare team
- Application of theoretical knowledge in practical scenarios
- Technical proficiency in respiratory care procedures

Utilization of Objective Measures: Participants apply objective measures and tools in competency evaluations

Overview: Participants will engage in applying objective measures and tools to assess competencies in respiratory care. The exercise focuses on utilizing standardized tests, skills checklists, and other objective assessment methods to evaluate theoretical knowledge and practical skills.

Activities:

1. *Standardized Knowledge Test*: Participants will take a standardized test covering theoretical aspects of respiratory care, including anatomy, physiology, and medical principles. The test will be structured to assess participants' knowledge-based competencies.
2. *Skills Checklists Assessment:* Participants will utilize skills checklists to evaluate specific hands-on skills and procedures. This activity will involve simulated scenarios where participants assess each other's proficiency in performing essential respiratory care procedures.
3. *Objective Structured Clinical Examination (OSCE): Participants* will engage in OSCE stations designed to assess comprehensive clinical competencies. Each station will focus on different aspects of clinical practice, and participants will use standardized checklists to evaluate their peers.

Assessment Criteria:

- Knowledge retention and application
- Proficiency in firsthand clinical skills
- Objective evaluation using standardized tools
- Holistic assessment of clinical competencies

Feedback and Reflection*: Discussion and reflection on the effectiveness of continuous competency assessment strategies*

Overview: This segment involves a group discussion and reflection on the continuous competency assessment strategies employed in respiratory care. Participants will share insights, discuss challenges, and reflect on the effectiveness of various assessment methods.

Discussion Points:

1. *Effectiveness of Scenario-Based Assessment:* Participants will discuss the realism and relevance of the simulated scenarios in assessing competencies. Insights into the challenges faced and lessons learned will be shared.
2. *Utilization of Objective Measures:* Participants will reflect on the strengths and limitations of using standardized tests, skills checklists, and OSCEs for competency evaluation. The discussion will explore the objectivity and reliability of these measures.
3. *Continuous Improvement Strategies:* The group will brainstorm strategies for continuous improvement in competency assessment, considering feedback received, emerging trends in respiratory care, and advancements in assessment methodologies.

Reflection Activities:

- Peer feedback on individual experiences
- Identification of areas for personal and collective improvement
- Consideration of innovative assessment approaches in respiratory care education

CHAPTER 4: CONSTRUCTIVE FEEDBACK TECHNIQUES

Objective:4

- Principles of delivering constructive feedback in precepting
- Feedback models and frameworks
- Addressing challenges and celebrating successes through feedback

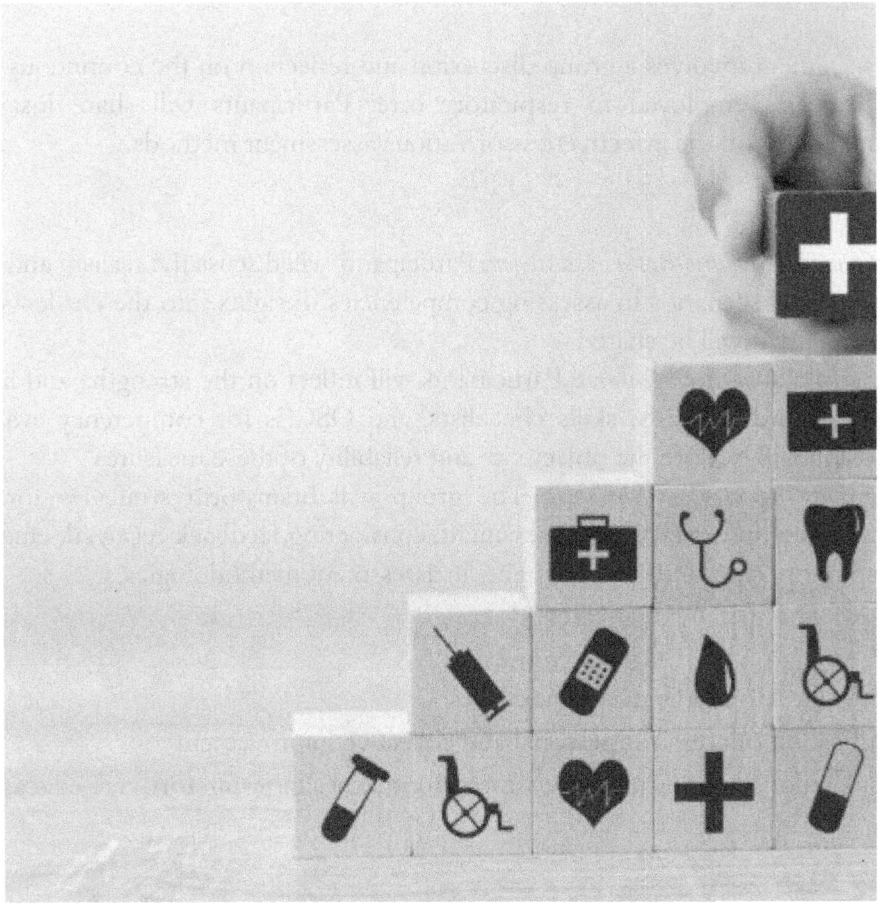

Principles of Constructive Feedback in Precepting

Overview: Fundamental principles of delivering constructive feedback in the context of respiratory care precepting. It emphasizes the importance of providing feedback that promotes learning, growth, and professional development.

Content:

1. *Timely Feedback:* The significance of delivering feedback in a timely manner to enhance its impact.
2. *Specificity and Clarity*: Principles of offering specific and clear feedback to address performance nuances.
3. *Positive and Constructive Tone:* Maintaining a positive and constructive tone to create a supportive learning environment.
4. *Individualized Feedback:* Recognizing the unique needs of each respiratory care student and tailoring feedback accordingly.

Timely Feedback: *The significance of delivering feedback in a timely manner to enhance its impact*

In respiratory care, timely feedback is crucial for maximizing its effectiveness. Providing feedback promptly after an event or performance allows students to connect it directly to their actions, enhancing understanding and retention. Timely feedback in respiratory care ensures that students can promptly address areas for improvement, reinforcing positive behaviors and facilitating a continuous learning cycle. This immediacy contributes to a dynamic and responsive educational environment, supporting the ongoing development of respiratory care skills.

Specificity and Clarity: *Principles of offering specific and clear feedback to address performance nuances*

In the context of respiratory care, specificity and clarity in feedback are essential for guiding students towards improvement. Clear feedback highlights specific actions, behaviors, or skills, enabling students to understand precisely what aspects of their performance are being addressed. Specific and clear feedback in respiratory care may involve detailing the correct execution of a procedure, pinpointing areas of improvement in patient assessment, or addressing technical skills. This approach fosters a targeted and constructive learning experience, facilitating the development of precise competencies in respiratory care.

Positive and Constructive Tone: *Maintaining a positive and constructive tone to create a supportive learning environment*

Maintaining a positive and constructive tone in feedback is integral to creating a supportive learning environment in respiratory care. Positive feedback reinforces strengths and achievements, motivating students to continue excelling. Constructive feedback, delivered in a supportive manner, focuses on areas for improvement without discouraging or demotivating students. In respiratory care, this approach fosters a culture of continuous improvement, where students feel encouraged to seek challenges, take risks, and engage in reflective practice to enhance their skills and knowledge.

Individualized Feedback: *Recognizing the unique needs of each respiratory care student and tailoring feedback accordingly*

Respiratory care students often have diverse learning styles, strengths, and areas for improvement. Individualized feedback acknowledges these differences and tailors the feedback process to meet the unique needs of each student. Recognizing individual strengths allows preceptors to reinforce positive behaviors, while addressing specific weaknesses ensures that feedback is relevant and impactful. Individualized feedback in respiratory care fosters a personalized learning experience, enhancing student engagement, motivation, and overall success in the field.

Feedback Models and Frameworks

Overview: Various feedback models and frameworks applicable to respiratory care precepting. It provides preceptors with structured approaches to delivering feedback effectively.

Content:

1. *The Pendleton Model:* Exploring the Pendleton model, which emphasizes a collaborative approach to feedback.
2. *Situational Leadership Feedback*: Applying situational leadership concepts to tailor feedback based on individual student needs.
3. *Feedback Sandwich Technique:* Understanding the feedback sandwich technique to balance positive and constructive feedback.
4. *360-Degree Feedback:* Exploring the use of feedback from multiple perspectives, including self-assessment, peer evaluation, and preceptor feedback.

The Pendleton Model: *Exploring the Pendleton model, which emphasizes a collaborative approach to feedback*

The Pendleton Model, particularly relevant in respiratory care, promotes a collaborative and learner-centered approach to feedback. In this model, the focus is on involving the learner in the feedback process actively. After the learner self-assesses, the preceptor provides positive aspects, followed by the learner's areas for improvement. The learner is then encouraged to identify their solutions and future strategies. This model empowers respiratory care students by actively engaging them in the feedback process, fostering self-directed learning, and promoting a sense of ownership over their professional development.

Situational Leadership Feedback: *Applying situational leadership concepts to tailor feedback based on individual student needs*

In the context of respiratory care, Situational Leadership Feedback adapts leadership concepts to the feedback process. This approach recognizes that different students may require varying types and levels of feedback based on their individual needs, competence, and readiness. Preceptors assess the specific situation and adapt their feedback style accordingly. For instance, a novice respiratory care student might benefit from more directive feedback, while an advanced student may respond well to a more collaborative and supportive approach. Situational Leadership Feedback ensures that feedback is tailored to meet the unique requirements of each student.

Feedback Sandwich Technique: *Understanding the feedback sandwich technique to balance positive and constructive feedback*

The Feedback Sandwich Technique is a widely used method in respiratory care education that involves sandwiching constructive feedback between positive feedback. In this approach, preceptors start with positive comments to reinforce what the student is doing well, then provide constructive feedback on areas for improvement, and conclude with additional positive reinforcement. This technique aims to maintain a positive and constructive tone, ensuring that students receive guidance on their strengths and areas for growth in a supportive manner. It helps create an atmosphere where students feel motivated to address challenges while recognizing their achievements.

360-Degree Feedback: *Exploring the use of feedback from multiple perspectives, including self-assessment, peer evaluation, and preceptor feedback*

In respiratory care, 360-Degree Feedback involves collecting feedback from various sources, providing a comprehensive view of a student's performance. This may include self-assessment, feedback from peers, and input from preceptors. This holistic approach allows students to gain insights into their strengths and areas for improvement from multiple perspectives. 360-Degree Feedback fosters a well-rounded understanding of a student's competencies, promoting a more thorough and accurate assessment of their abilities in respiratory care practice.

Exercise: *Applying Feedback Techniques in Precepting*

Overview: This practical exercise focuses on applying feedback techniques in the precepting context. Participants will engage in simulated scenarios and discussions to practice and refine their feedback delivery skills.

Exercise Content:

1. *Role-Playing Scenarios:* Participants will participate in role-playing scenarios where they take turns providing feedback to simulated respiratory care students.

2. *Feedback Roundtable Discussion:* A structured discussion on the challenges and successes experienced during the role-playing exercises.
3. *Reflection and Improvement Plans*: Participants will reflect on their feedback delivery and collaboratively develop improvement plans.

Exercise Content for Role-Playing Scenarios: *Participants will participate in role-playing scenarios where they take turns providing feedback to simulated respiratory care students*

Overview: This exercise immerses participants in realistic role-playing scenarios, allowing them to practice delivering feedback to simulated respiratory care students. The focus is on applying feedback models and frameworks learned during the lectures.

Activities:

1. *Simulated Scenarios:* Participants will rotate through various simulated scenarios, each depicting common situations in respiratory care practice.
2. *Feedback Delivery:* Participants will take turns playing the role of a preceptor providing feedback to a simulated respiratory care student. Feedback should align with principles of constructive feedback, incorporating specific and timely elements.

Assessment Criteria:

- Application of feedback models (e.g., Pendleton Model, Feedback Sandwich Technique)
- Clarity and specificity in feedback delivery
- Use of positive and constructive tone
- Ability to tailor feedback to the simulated student's level and needs

Feedback Roundtable Discussion: *A structured discussion on the challenges and successes experienced during the role-playing exercises*

Overview: After the role-playing scenarios, participants engage in a structured roundtable discussion to share their experiences, challenges, and successes encountered while delivering feedback. The aim is to promote collective learning and identify areas for improvement.

Discussion Points:

1. *Challenges Faced:* Participants discuss challenges encountered during the role-playing exercises, such as delivering difficult feedback or adapting feedback to different scenarios.
2. *Successes and Effective Strategies:* Participants share successful feedback delivery strategies, highlighting approaches that were particularly effective in the simulated scenarios.

3. *Peer Insights:* Participants provide constructive feedback to each other on their feedback delivery, fostering a collaborative learning environment.

Reflection and Learning Objectives:

- Recognizing familiar challenges in delivering feedback in respiratory care.
- Identifying effective strategies for overcoming feedback delivery challenges.
- Enhancing peer-to-peer feedback and collaborative learning.

Reflection and Improvement Plans: *Participants will reflect on their feedback delivery and collaboratively develop improvement plans*

Overview: This segment focuses on individual and collective reflection. Participants will reflect on their feedback delivery experiences, identify areas for improvement, and collaboratively develop plans to enhance their feedback skills in respiratory care precepting.

Activities:

1. *Individual Reflection:* Participants reflect on their firsthand experiences during the role-playing scenarios, considering strengths and areas for improvement in feedback delivery.
2. *Group Discussion*: Participants share reflections with the group, discussing common themes and insights.
3. *Collaborative Improvement Plans:* In small groups, participants collaborate to develop improvement plans, outlining specific actions to enhance their feedback proficiency.

Learning Objectives:

- Self-awareness of feedback delivery strengths and areas for improvement.
- Collaboration in identifying common improvement themes.
- Development of actionable improvement plans for feedback proficiency.

This exercise aims to provide participants with practical experience, foster collaborative learning, and promote continuous improvement in delivering constructive feedback in respiratory care precepting.

Chapter 5: Individualized Learning Plans

Objective:5

- Designing personalized learning plans based on individual student needs
- Accommodating diverse learning styles in respiratory care education
- Adapting learning plans for optimal student development

Designing Personalized Learning Plans

Overview: The principles and strategies for designing personalized learning plans in respiratory care education. It emphasizes the importance of tailoring educational experiences to meet the individual needs, goals, and learning preferences of respiratory care students.

Content:

1. *Assessment of Individual Needs:* Techniques for identifying the unique learning needs and preferences of respiratory care students.
2. *Setting Measurable Goals:* Establishing clear and measurable learning objectives that align with the individual's educational and professional aspirations.
3. *Selecting Appropriate Learning Activities:* Choosing diverse and effective learning activities that cater to different learning styles and preferences.
4. *Feedback and Reflection:* Incorporating mechanisms for ongoing feedback and reflection to adapt learning plans dynamically.

Assessment of Individual Needs: *Techniques for identifying the unique learning needs and preferences of respiratory care students*

In the context of respiratory care education, assessment of individual needs involves employing various techniques to understand the specific learning requirements and preferences of each student. This process may include surveys, interviews, and observation to gather information about students' prior knowledge, learning styles, and areas of strength or challenge. By comprehensively assessing individual needs, educators in respiratory care can tailor their instructional approaches, ensuring that learning plans are personalized and responsive to the diverse characteristics of each student.

Setting Measurable Goals: Establishing clear and measurable learning objectives that align with the individual's educational and professional aspirations

Setting measurable goals in respiratory care education involves defining clear and specific learning objectives that align with each student's educational and professional aspirations. These goals are concrete, observable, and quantifiable, allowing both students and educators to track progress effectively. In respiratory care, measurable goals may include mastering specific clinical skills, achieving proficiency in diagnostic procedures, or attaining a certain level of competence in patient care. By aligning goals with individual aspirations, educators facilitate meaningful and targeted learning experiences, fostering motivation and success.

Selecting Appropriate Learning Activities: *Choosing diverse and effective learning activities that cater to different learning styles and preferences*

In respiratory care education, selecting appropriate learning activities entails choosing a variety of instructional methods that cater to diverse learning styles and preferences. This may involve incorporating visual aids, hands-on simulations, interactive discussions, and case studies. Recognizing that respiratory care students may have different learning preferences—such as visual, auditory, kinesthetic, or read/write—educators aim to create a well-rounded curriculum that engages students effectively. This approach enhances the overall learning experience, ensuring that each student can grasp and apply respiratory care concepts in ways that resonate with their individual learning style.

Feedback and Reflection: *Incorporating mechanisms for ongoing feedback and reflection to adapt learning plans dynamically*

Feedback and reflection play a crucial role in respiratory care education by providing mechanisms for ongoing assessment and adaptation of learning plans. Educators use feedback to inform students about their progress, strengths, and areas for improvement. Regular reflection allows students to assess their own learning experiences and make adjustments accordingly. Dynamic adaptation of learning plans involves responding to feedback and reflections by modifying instructional strategies, content delivery, or assessment methods. This iterative process ensures that respiratory care education remains responsive to the evolving needs and progress of individual students, fostering continuous improvement and personalized development.

Accommodating Diverse Learning Styles

Overview: The concept of diverse learning styles in respiratory care education and provides insights into accommodating these styles effectively. Understanding and addressing diverse learning preferences contribute to the development of inclusive and effective learning plans.

Content:

1. *Identifying Learning Style Preferences:* Recognizing and assessing various learning style preferences among respiratory care students.
2. *Adapting Instructional Strategies:* Tailoring instructional strategies to align with different learning styles, including visual, auditory, kinesthetic, and read/write.
3. *Creating Multimodal Learning Experiences:* Incorporating a variety of learning modalities to engage students with diverse preferences.
4. *Promoting Active Learning:* Encouraging active participation and interactive elements in learning plans to cater to different learning styles.

Identifying Learning Style Preferences: *Recognizing and assessing various learning style preferences among respiratory care students*

Identifying learning style preferences in respiratory care education involves recognizing and assessing the diverse ways in which students prefer to acquire and process information. Learning styles may encompass visual, auditory, kinesthetic, or read/write preferences. Educators use tools such as surveys, observations, or discussions to identify individual learning styles. Recognizing these preferences allows educators in respiratory care to tailor their instructional approaches, ensuring that the content is presented in ways that resonate with each student's preferred mode of learning.

Adapting Instructional Strategies: *Tailoring instructional strategies to align with different learning styles, including visual, auditory, kinesthetic, and read/write*

Adapting instructional strategies in respiratory care education entails tailoring teaching methods to align with the identified learning styles of students. For visual learners, educators might incorporate diagrams, charts, or videos. Auditory learners may benefit from lectures, discussions, or recorded materials. Kinesthetic learners thrive with hands-on activities and simulations, while read/write learners prefer written materials and note-taking. By adapting instructional strategies, educators cater to the diverse needs of respiratory care students, enhancing comprehension and engagement.

Creating Multimodal Learning Experiences: *Incorporating a variety of learning modalities to engage students with diverse preferences*

Creating multimodal learning experiences in respiratory care involves integrating a variety of instructional modalities within a single learning environment. This approach recognizes that students may have multiple learning style preferences and benefits from exposure to diverse teaching methods. Educators might combine visual presentations with interactive discussions, hands-on simulations, and written assignments. Multimodal learning experiences provide respiratory care students with a holistic and comprehensive educational journey, accommodating various preferences and optimizing the learning process.

Promoting Active Learning: *Encouraging active participation and interactive elements in learning plans to cater to different learning styles*

Promoting active learning in respiratory care education involves encouraging students to actively engage with the learning process. This includes interactive discussions, group activities, case studies, and hands-on simulations. Active learning approaches accommodate different learning styles by providing opportunities for participation, collaboration, and application of knowledge. By incorporating interactive elements, educators foster a dynamic and engaging learning environment that aligns with the diverse preferences of respiratory care students, promoting deeper understanding and retention of content.

Practical Session: *Adapting Learning Plans for Student Development*

Overview: This practical session involves hands-on activities where participants apply the concepts from the lectures to adapt and customize learning plans for respiratory care students. The focus is on incorporating individualized elements based on needs, goals, and learning styles.

Session Content:

1. *Case Studies:* Participants analyze case studies representing diverse respiratory care students, identifying their unique needs and preferences.
2. *Group Work*: Small groups collaborate to adapt generic learning plans to address the individualized requirements of specific cases.
3. *Peer Review:* Groups present their adapted learning plans, and participants engage in peer review discussions to provide constructive feedback.
4. *Reflection and Iteration:* Participants reflect on the effectiveness of their adaptations and discuss potential iterations for further improvement.

Session Content: *Adaptation of Learning Plans for Respiratory Care Students*

Overview: This session is designed to engage participants in practical activities that involve adapting learning plans based on case studies representing diverse respiratory care students. The focus is on recognizing individual needs and preferences to create personalized and effective learning experiences.

Activities:

1. **Case Studies:**
   a. Participants are presented with case studies that portray different respiratory care students, each with unique learning needs, preferences, and challenges.
   b. The case studies may include scenarios such as students with varied learning styles, academic backgrounds, or clinical experiences.
   c. Participants analyze the case studies to identify the individualized requirements of each student.
2. **Group Work:**
   a. Participants form small groups, and each group is assigned a specific case study.
   b. Groups collaborate to adapt generic learning plans to address the individualized needs, preferences, and challenges outlined in their assigned case study.
   c. The adaptation process involves selecting appropriate instructional strategies, setting measurable goals, and incorporating diverse learning modalities.
3. **Peer Review:**
   a. Groups present their adapted learning plans to the entire session participants.
   b. After each presentation, participants engage in peer review discussions, providing constructive feedback on the effectiveness and appropriateness of the adaptations.

   c. Peer review focuses on alignment with individual needs, creativity in adapting strategies, and the feasibility of implementation.

4. **Reflection and Iteration:**
   a. Participants engage in a reflective discussion on the adaptation process, considering challenges encountered and successes achieved.
   b. Reflection includes an assessment of the effectiveness of the adapted learning plans in meeting the unique needs of respiratory care students.
   c. Participants discuss potential iterations and improvements for further enhancing the individualization of learning plans.

Learning Objectives:

- Recognition of diverse learning needs and preferences in respiratory care students.
- Application of adaptive strategies to create personalized learning plans.
- Development of peer-review skills for constructive feedback.
- Reflection on the iterative process of continuous improvement in adapting learning plans.

CHAPTER 6: ETHICAL CONSIDERATIONS IN PRECEPTING

Objective: 6

- Identifying and navigating ethical dilemmas in precepting
- Establishing a culture of professionalism and integrity
- Ethical decision-making in respiratory care education

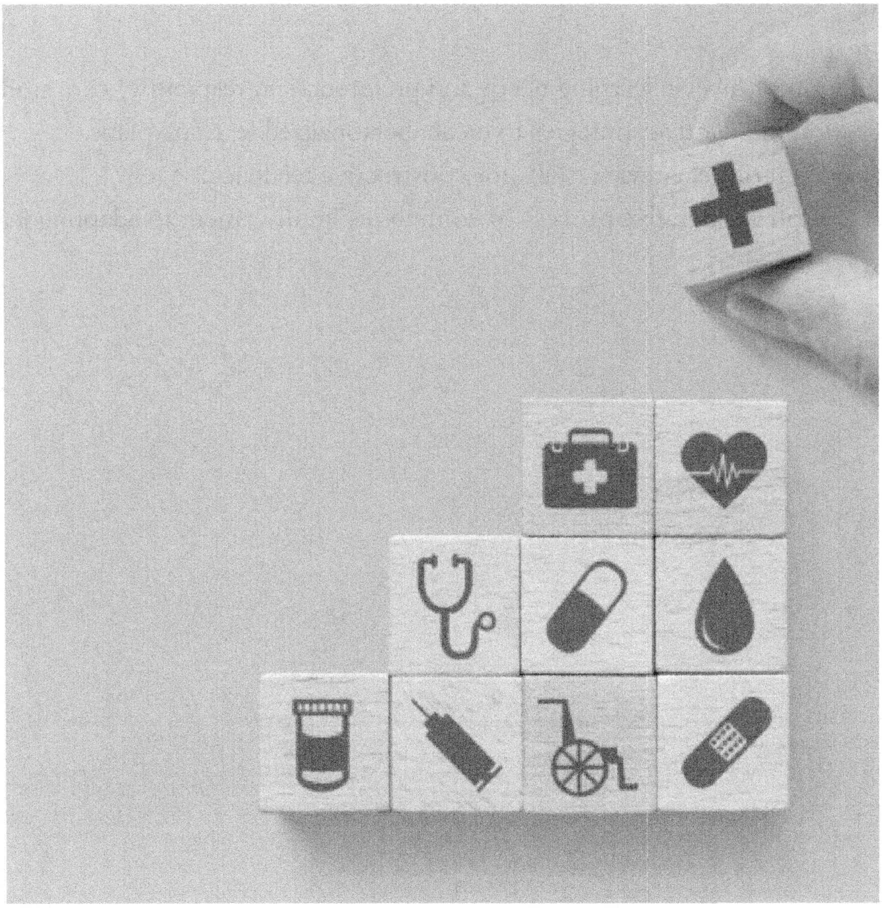

Identifying Ethical Dilemmas in Precepting

Overview: Recognizing and navigating ethical dilemmas that may arise in the precepting context of respiratory care education. Emphasis is placed on developing awareness, critical thinking, and ethical decision-making skills to address complex situations.

Content:

1. *Types of Ethical Dilemmas:* Exploring common ethical dilemmas encountered in respiratory care precepting, such as issues related to student supervision, evaluation, and professional boundaries.
2. *Ethical Decision-Making Models:* Introducing ethical decision-making models that can guide preceptors in systematically addressing ethical challenges.
3. *Communication Strategies:* Discussing effective communication strategies for addressing ethical concerns with students, colleagues, and other healthcare professionals.
4. *Legal and Ethical Considerations:* Highlighting the intersection of legal and ethical considerations in precepting, emphasizing the importance of compliance with professional standards.

Types of Ethical Dilemmas: *Exploring common ethical dilemmas encountered in respiratory care precepting, such as issues related to student supervision, evaluation, and professional boundaries*

In respiratory care precepting, ethical dilemmas often arise in various aspects of student mentorship. Common dilemmas may involve issues related to the supervision and evaluation of students, where preceptors must navigate challenges, such as providing constructive feedback, ensuring fair evaluation, and addressing academic or clinical performance concerns. Professional boundaries, another critical area, may present dilemmas related to maintaining appropriate relationships, confidentiality, and avoiding conflicts of interest. Recognizing these dilemmas is essential for preceptors to approach ethical decision-making with a comprehensive understanding of the complexities involved in guiding and assessing respiratory care students.

Ethical Decision-Making Models: *Introducing ethical decision-making models that can guide preceptors in systematically addressing ethical challenges*

To navigate ethical dilemmas effectively, preceptors in respiratory care benefit from employing ethical decision-making models. These models provide a systematic framework for analyzing situations, considering various perspectives, and arriving at ethically sound decisions. Ethical decision-making models typically involve steps such as identifying the issue, gathering relevant information, considering ethical principles, exploring alternative courses of action, and making a reasoned decision. By introducing and applying these models, preceptors enhance their ability to approach ethical challenges in a structured and principled manner, fostering a culture of ethical conduct in respiratory care education.

Communication Strategies: *Discussing effective communication strategies for addressing ethical concerns with students, colleagues, and other healthcare professionals*

Effective communication is paramount when addressing ethical concerns in respiratory care precepting. Preceptors need to engage in open and transparent communication with students, colleagues, and other healthcare professionals. This involves expressing concerns clearly, actively listening to different perspectives, and collaboratively working towards resolutions. Communication strategies also include fostering an environment where students feel comfortable discussing ethical issues, seeking guidance, and participating in ethical decision-making processes. Effective communication contributes to a supportive and ethical learning environment in respiratory care education.

Legal and Ethical Considerations: *Highlighting the intersection of legal and ethical considerations in precepting, emphasizing the importance of compliance with professional standards*

The intersection of legal and ethical considerations is a crucial aspect of precepting in respiratory care. Preceptors must be aware of both legal requirements and ethical principles governing their practice. This involves compliance with professional standards, codes of ethics, and relevant laws in respiratory care education. Highlighting this intersection emphasizes the importance of aligning ethical conduct with legal obligations. Preceptors need to navigate situations where legal and ethical considerations may converge or diverge, ensuring that their actions adhere to both the letter of the law and the spirit of ethical practice in respiratory care. This awareness is vital for promoting integrity and accountability in precepting roles.

Establishing a Culture of Professionalism Case Studies: *Ethical Decision-Making in Respiratory Care Education*

Overview: The importance of fostering a culture of professionalism in respiratory care education. Case studies are used to illustrate real-world scenarios, prompting participants to engage in ethical decision-making and exploring strategies for establishing and maintaining a professional learning environment.

Content:

1. *Professionalism in Respiratory Care:* Defining and emphasizing the key components of professionalism within the context of respiratory care education.
2. *Case Studies:* Presenting case studies that depict ethical dilemmas related to professionalism, encouraging participants to analyze, discuss, and apply ethical decision-making principles.

3. *Establishing Expectations:* Discussing the role of preceptors in setting clear expectations for ethical behavior and professionalism among respiratory care students.
4. *Cultivating a Supportive Environment:* Exploring strategies for creating an environment that supports open communication, ethical reflection, and continuous improvement.

Professionalism in Respiratory Care: *Defining and emphasizing the key components of professionalism within the context of respiratory care education*

Professionalism in respiratory care education encompasses a set of values, behaviors, and attitudes that reflect the high standards expected of practitioners in the field. This involves adhering to ethical principles, maintaining competence, and demonstrating a commitment to patient-centered care. Within the educational context, professionalism extends to how preceptors guide and assess students, emphasizing the importance of ethical conduct, responsibility, and respect for colleagues and patients. Defining and emphasizing key components of professionalism provides a foundation for ethical and accountable practice in respiratory care education.

Case Studies: *Presenting case studies that depict ethical dilemmas related to professionalism, encouraging participants to analyze, discuss, and apply ethical decision-making principles*

Case studies serve as valuable tools for exploring ethical dilemmas related to professionalism in respiratory care education. By presenting real-world scenarios, participants are prompted to analyze, discuss, and apply ethical decision-making principles to resolve complex situations. These case studies may involve issues such as conflicts of interest, breaches of confidentiality, or challenging interactions with colleagues. Engaging with these cases encourages participants to develop critical thinking skills, consider various perspectives, and apply ethical principles to make informed decisions in the realm of respiratory care professionalism.

Establishing Expectations: *Discussing the role of preceptors in setting clear expectations for ethical behavior and professionalism among respiratory care students*

Preceptors play a crucial role in shaping the professional development of respiratory care students. Establishing clear expectations for ethical behavior and professionalism is essential in guiding students toward the high standards expected in the field. Discussions on this topic involve outlining professional responsibilities, ethical considerations, and the importance of maintaining a professional demeanor in educational and clinical settings. Clear expectations provide a framework for students to understand the behaviors and attitudes required for success in respiratory care, fostering a culture of professionalism.

Cultivating a Supportive Environment: *Exploring strategies for creating an environment that supports open communication, ethical reflection, and continuous improvement*

Cultivating a supportive environment is integral to promoting professionalism in respiratory care education. This involves creating a culture where open communication is encouraged, allowing students and preceptors to discuss ethical concerns openly. Strategies for ethical reflection, such as regular debriefings or reflective exercises, contribute to continuous improvement and the development of a strong ethical foundation. A supportive environment ensures that challenges are addressed collaboratively, fostering a culture of learning, accountability, and professionalism in the respiratory care education community.

Chapter 7: Cultural Competence in Precepting

Objective:7

- Understanding cultural competence in healthcare education
- Creating an inclusive and culturally sensitive learning environment
- Addressing cultural diversity challenges in precepting

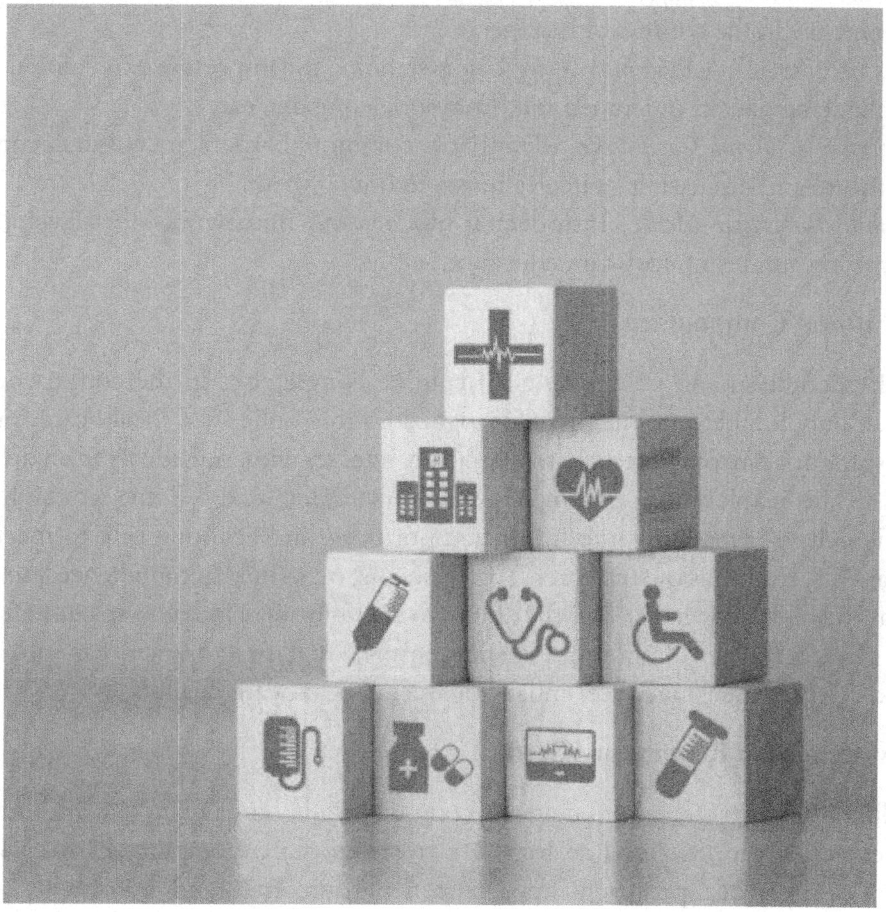

Understanding Cultural Competence in Healthcare

Overview: The significance of cultural competence in healthcare, emphasizing its role in respiratory care education. The aim is to provide preceptors with an understanding of cultural competence, its impact on patient care, and strategies for integrating cultural awareness into educational practices.

Content:

1. *Defining Cultural Competence:* Exploring the definition and components of cultural competence in the context of healthcare.
2. *Cultural Awareness in Respiratory Care*: Understanding the importance of cultural competence in enhancing patient outcomes and promoting equitable care.
3. *Barriers to Cultural Competence*: Identifying common barriers to cultural competence in healthcare and discussing strategies for overcoming them.
4. *Cultural Competence Models:* Introducing models and frameworks for developing cultural competence in respiratory care education.

Defining Cultural Competence

Exploring the definition and components of cultural competence in the context of healthcare: Defining cultural competence involves recognizing the ability of healthcare professionals, including respiratory care practitioners, to effectively interact with individuals from diverse cultural backgrounds. This exploration encompasses understanding and valuing cultural differences, incorporating cultural awareness into healthcare practices, and adapting care to meet the unique needs of patients from various cultures. Components of cultural competence include cultural sensitivity, awareness, knowledge, and the ability to skillfully navigate diverse cultural contexts. In respiratory care, defining cultural competence ensures that practitioners can provide patient-centered and culturally sensitive care, enhancing overall healthcare quality.

Cultural Awareness in Respiratory Care

Understanding the importance of cultural competence in enhancing patient outcomes and promoting equitable care: Cultural awareness in respiratory care emphasizes the recognition of how cultural competence positively impacts patient outcomes and contributes to equitable healthcare. Understanding diverse cultural backgrounds enables respiratory care practitioners to deliver care that is respectful, responsive, and tailored to individual needs. Cultural awareness involves recognizing the influence of cultural factors on health beliefs, communication styles, and treatment preferences. By prioritizing cultural awareness, respiratory care practitioners contribute to improved patient satisfaction, adherence to treatment plans, and overall health outcomes.

Barriers to Cultural Competence

Identifying common barriers to cultural competence in healthcare and discussing strategies for overcoming them: Barriers to cultural competence in healthcare refer to challenges that may impede effective interactions with patients from diverse cultural backgrounds. Common barriers include language barriers, lack of cultural awareness among healthcare providers, and stereotyping. Identifying these barriers is essential for developing strategies to overcome them. Discussions may focus on promoting cultural competency training, implementing interpreter services, and fostering a healthcare environment that values diversity. Overcoming these barriers is crucial for creating an inclusive and culturally competent respiratory care practice.

Cultural Competence Models

Introducing models and frameworks for developing cultural competence in respiratory care education: Introducing cultural competence models involves presenting structured frameworks and approaches that can guide the development of cultural competence in respiratory care education. These models may include stages of cultural competence, competency domains, or systematic processes for integrating cultural awareness into educational curricula. By introducing these models, respiratory care educators can provide students with a structured pathway to acquiring cultural competence skills, ultimately preparing them to deliver high-quality and culturally sensitive care in their professional practice.

Creating an Inclusive Learning Environment

Overview: Creating an inclusive learning environment in respiratory care education, where cultural competence is integrated seamlessly. Participants will learn strategies to foster a climate that respects diversity, promotes inclusivity, and supports students from various cultural backgrounds.

Content:

1. *Inclusive Pedagogy:* Discussing teaching strategies that accommodate diverse learning styles and cultural backgrounds.
2. *Curriculum Development:* Integrating cultural competence into the respiratory care curriculum to reflect a diverse range of patient populations.
3. *Cultural Competence Assessment:* Exploring methods for assessing and evaluating cultural competence among respiratory care students.
4. *Promoting Cultural Humility:* Emphasizing the concept of cultural humility and its application in respiratory care education.

Inclusive Pedagogy: *Discussing teaching strategies that accommodate diverse learning styles and cultural backgrounds*

Inclusive pedagogy involves the deliberate design and implementation of teaching strategies that embrace and accommodate diverse learning styles and cultural backgrounds. In respiratory care education, this means recognizing that students may have different ways of processing information and diverse cultural influences that shape their learning preferences. Discussions on inclusive pedagogy explore techniques such as varied instructional methods, flexible assessment approaches, and creating an environment that fosters engagement and participation for all students. By incorporating inclusive pedagogy, respiratory care educators enhance the accessibility of their teaching methods, ensuring that students from diverse backgrounds can thrive in the learning environment.

Curriculum Development: *Integrating cultural competence into the respiratory care curriculum to reflect a diverse range of patient populations*

Curriculum development in respiratory care education involves intentionally integrating cultural competence throughout the educational program. This includes incorporating content that reflects a diverse range of patient populations, cultural contexts, and healthcare disparities. By infusing cultural competence into the curriculum, students gain a deeper understanding of the impact of culture on healthcare practices. Curriculum development discussions explore ways to address cultural competence within core courses, clinical experiences, and simulations. This integration ensures that respiratory care students are well-prepared to provide inclusive and culturally sensitive care in their future professional roles.

Cultural Competence Assessment: *Exploring methods for assessing and evaluating cultural competence among respiratory care students*

Cultural competence assessment focuses on evaluating the ability of respiratory care students to effectively navigate and respond to cultural diversity in healthcare settings. Discussions on this topic explore various assessment methods, such as written assignments, case studies, simulations, and reflective exercises, to measure students' cultural competence. The goal is to ensure that students can apply theoretical knowledge into practical skills and demonstrate cultural sensitivity in their interactions with patients. Cultural competence assessment is an integral part of evaluating the effectiveness of cultural competence education in respiratory care programs.

Promoting Cultural Humility: *Emphasizing the concept of cultural humility and its application in respiratory care education*

Promoting cultural humility involves emphasizing a lifelong commitment to self-reflection, learning, and openness to different cultural perspectives. In respiratory care education, this concept encourages students to approach cultural competence with humility, acknowledging that their understanding is an ongoing process. Discussions on promoting cultural humility explore ways to foster a mindset of continuous learning, self-awareness, and responsiveness to the unique needs and values of diverse patient populations. By incorporating cultural humility into respiratory care education, students are better equipped to navigate the complexities of cultural diversity in healthcare with sensitivity and respect.

Group Discussion: *Addressing Cultural Diversity Challenges*

Overview: This interactive session encourages participants to engage in a group discussion format, addressing challenges related to cultural diversity in respiratory care education. The goal is to facilitate open dialogue, share experiences, and collaboratively identify strategies to overcome cultural diversity challenges.

Discussion Topics:

1. *Personal Experiences:* Participants share personal experiences related to cultural diversity challenges in respiratory care education.
2. *Identifying Common Challenges:* Group discussion on common challenges faced by preceptors and students in a culturally diverse learning environment.
3. *Strategies for Overcoming Challenges:* Collaborative exploration of effective strategies to overcome cultural diversity challenges, drawing on collective insights.
4. *Role-Playing Scenarios*: Practical exercises involving role-playing scenarios to simulate and address cultural diversity challenges.

Discussion Topics: Cultural Diversity Challenges in Respiratory Care Education

1. Personal Experiences:

- Participants share firsthand experiences related to cultural diversity challenges in respiratory care education.
- Each participant has the opportunity to discuss instances where they encountered cultural diversity challenges, reflecting on the impact on teaching or learning experiences.
- Emphasis on sharing insights gained, lessons learned, and personal growth in navigating cultural diversity challenges.

2. Identifying Common Challenges:

- Group discussion on common challenges faced by preceptors and students in a culturally diverse learning environment.
- Participants collaboratively identify recurring themes and patterns related to cultural diversity challenges, acknowledging shared experiences.
- Exploration of challenges may include communication barriers, differing cultural norms, and potential misunderstandings.

3. Strategies for Overcoming Challenges:

- Collaborative exploration of effective strategies to overcome cultural diversity challenges, drawing on collective insights.
- Participants contribute ideas and best practices for fostering an inclusive and culturally sensitive learning environment.
- Strategies may encompass initiative-taking communication, culturally tailored instructional methods, and interventions to address cultural competence gaps.

4. Role-Playing Scenarios:

- Practical exercises involving role-playing scenarios to simulate and address cultural diversity challenges.
- Participants actively engage in role-playing activities that mirror real-life situations, allowing them to apply cultural competence strategies in a controlled environment.
- Discussion follows each scenario, enabling participants to analyze outcomes, share alternative approaches, and refine their cultural competence skills.

Overall Learning Objectives:

- Enhance self-awareness and reflection on personal experiences with cultural diversity challenges.
- Identify common challenges faced by preceptors and students in respiratory care education.
- Collaboratively develop effective strategies for overcoming cultural diversity challenges.
- Apply cultural competence skills through role-playing scenarios to simulate and address real-world challenges in respiratory care education.

MODULE 8: EVIDENCE-BASED PRACTICES IN PRECEPTING

Objective:8

- Overview of evidence-based practices in respiratory care education
- Integration of research findings into precepting strategies
- Staying informed about the latest research in the field

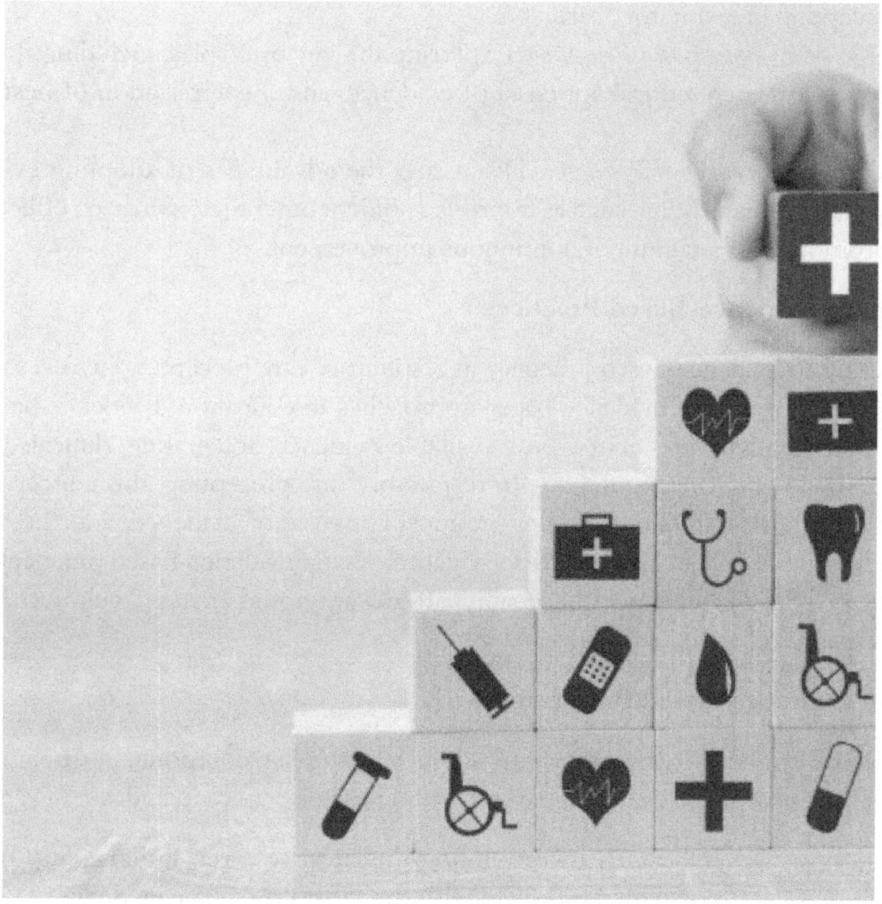

Overview of Evidence-Based Practices

Overview: This introduction to evidence-based practices in precepting within the context of respiratory care education. The aim is to equip preceptors with an understanding of the principles of evidence-based precepting and its significance in enhancing the quality of education and patient care.

Content:

1. *Introduction to Evidence-Based Practices*: Defining evidence-based practices and their relevance to precepting in respiratory care.
2. *Principles of Evidence-Based Precepting:* Exploring the key principles, including the integration of current research, critical appraisal of evidence, and the application of best practices in educational settings.
3. *Benefits of Evidence-Based Precepting:* Discussing the advantages of adopting evidence-based approaches in precepting, such as improved student outcomes, increased effectiveness, and the promotion of a culture of continuous improvement.

Introduction to Evidence-Based Practices

The introduction to evidence-based practices in respiratory care precepting involves defining and emphasizing the relevance of evidence-based approaches in education. Evidence-based practices refer to the conscientious use of the best available evidence, integrating clinical expertise and patient values to inform decision-making. In respiratory care precepting, this concept emphasizes the application of current research findings, critical appraisal of evidence, and the alignment of teaching methods with proven best practices. By introducing evidence-based practices, preceptors in respiratory care aim to enhance the quality of education and foster a culture of continuous improvement based on reliable evidence.

Principles of Evidence-Based Precepting

The principles of evidence-based precepting delve into the key components that guide effective and informed teaching practices in respiratory care. These principles include:

1. *Integration of Current Research*: Incorporating the latest research findings into precepting to ensure that educational methods align with the most up-to-date evidence.
2. *Critical Appraisal of Evidence*: Developing the skills to critically appraise research evidence relevant to respiratory care education, enabling preceptors to discern the validity and applicability of different studies.
3. *Application of Best Practices:* Applying evidence-based best practices in educational settings, aligning teaching methods with proven approaches to optimize student learning outcomes.

These principles guide preceptors in making informed decisions, adapting their teaching strategies based on the latest evidence, and fostering a dynamic and evidence-informed educational environment.

Benefits of Evidence-Based Precepting

Discussing the benefits of evidence-based precepting highlights the positive outcomes associated with adopting evidence-informed approaches in respiratory care education. Some advantages include:

1. *Improved Student Outcomes:* Evidence-based precepting contributes to enhanced student learning experiences and outcomes by aligning teaching methods with proven effective strategies.
2. *Increased Effectiveness:* Preceptors who incorporate evidence-based practices are more likely to be effective in their teaching, as they utilize the best available evidence to inform their decisions and approaches.
3. *Promotion of Continuous Improvement:* Embracing evidence-based precepting fosters a culture of continuous improvement, encouraging preceptors to regularly update their practices based on the latest research and strive for excellence in respiratory care education.

Integrating Research Findings into Precepting

Overview: The practical aspects of integrating research findings into the precepting process. It aims to empower preceptors with the skills to critically appraise research, apply evidence to their teaching methods, and enhance the overall educational experience for respiratory care students.

Content:

1. *Critical Appraisal of Research:* Providing preceptors with tools and techniques for critically appraising research findings relevant to respiratory care education.
2. *Application of Evidence in Teaching:* Exploring strategies for incorporating research evidence into precepting practices, including adapting teaching methods based on current research.
3. *Case Studies:* Applying Research in Precepting: Presenting case studies that demonstrate the effective integration of research findings into precepting scenarios, highlighting successful approaches.

Critical Appraisal of Research

Providing preceptors with tools and techniques for critically appraising research findings relevant to respiratory care education: Critical appraisal of research involves equipping preceptors with the skills to assess and evaluate research findings in the context of respiratory care education. This includes providing tools and techniques to systematically evaluate the quality, relevance, and applicability of research studies. Preceptors learn to discern the strengths and limitations of research evidence, enabling them to make informed decisions about the integration of research findings into their teaching practices. The goal is to empower preceptors to engage with research literature in a critical and meaningful way, ensuring that evidence used in precepting is of high quality and aligns with educational goals.

Application of Evidence in Teaching

Exploring strategies for incorporating research evidence into precepting practices, including adapting teaching methods based on current research: The application of evidence in teaching focuses on practical strategies for integrating research findings into precepting practices in respiratory care education. Preceptors explore ways to adapt their teaching methods based on current research, ensuring that educational approaches align with evidence-based best practices. This includes incorporating innovative teaching strategies supported by research, updating instructional materials considering new evidence, and adjusting the curriculum to reflect the latest advancements in respiratory care. By actively applying evidence in teaching, preceptors enhance the relevance and effectiveness of their educational methods.

Case Studies

Applying Research in Precepting: Presenting case studies that demonstrate the effective integration of research findings into precepting scenarios, highlighting successful approaches: The use of case studies involves presenting real-world scenarios that demonstrate how research findings can be effectively integrated into precepting in respiratory care education. These case studies showcase successful approaches, illustrating how evidence-based practices can be applied in diverse precepting situations. Preceptors engage in discussions around these case studies, analyzing the decision-making processes, identifying key strategies, and reflecting on the outcomes. Case studies provide practical examples that enhance preceptors' understanding of the application of research evidence in precepting, fostering a deeper appreciation for evidence-based approaches.

In summary, critical appraisal of research, application of evidence in teaching, and case studies collectively contribute to the development of preceptors' skills in effectively utilizing research evidence in respiratory care education. These approaches ensure that precepting practices are grounded in high-quality evidence, promoting continuous improvement and innovation in teaching methods.

Reading Assignment: *Staying Informed about Latest Research*

Overview: This reading assignment focuses on the importance of staying informed about the latest research in respiratory care and education. It encourages preceptors to engage in ongoing learning and continuous professional development by regularly reviewing current literature.

Content:

1. *Navigating Respiratory Care Journals:* Guidance on accessing and navigating reputable respiratory care journals to stay updated on the latest research.
2. *Database Search Strategies:* Tips for effective database searches to identify relevant research articles and educational resources.
3. *Critical Reading Skills:* Developing critical reading skills to evaluate the quality and applicability of research findings.
4. *Discussion on Selected Articles:* Participants engage in discussions on selected research articles related to respiratory care education.

Navigating Respiratory Care Journals: *Guidance on accessing and navigating reputable respiratory care journals to stay updated on the latest research*

Navigating respiratory care journals involves providing preceptors with guidance on accessing and exploring reputable journals specific to respiratory care. This includes introducing preceptors to key journals in the field, highlighting their significance, and demonstrating effective ways to navigate and stay updated on the latest research. By understanding how to access and explore relevant journals, preceptors can regularly engage with current literature, ensuring they are informed about advancements and research findings in respiratory care education.

Database Search Strategies: *Tips for effective database searches to identify relevant research articles and educational resources*

Database search strategies equip preceptors with the skills to conduct effective searches to identify pertinent research articles and educational resources in respiratory care. This involves providing tips on using keywords, Boolean operators, and filters to streamline searches and retrieve the most relevant information. Effective database search strategies enable preceptors to efficiently locate evidence that aligns with their specific areas of interest or teaching needs. This skill is crucial for staying informed about the latest research and educational developments in respiratory care.

Critical Reading Skills: *Developing critical reading skills to evaluate the quality and applicability of research findings*

Critical reading skills are essential for preceptors to evaluate the quality and applicability of research findings in respiratory care. This involves teaching preceptors how to critically assess research articles, considering factors such as study design, sample size, methodology, and statistical analysis. Developing these skills enables preceptors to discern the validity and relevance of research evidence, facilitating informed decision-making when integrating evidence into their precepting practices.

Discussion on Selected Articles: *Participants engage in discussions on selected research articles related to respiratory care education*

Engaging in discussions on selected articles fosters a collaborative learning environment where preceptors can share insights, perspectives, and reflections on research relevant to respiratory care education. This activity involves selecting specific articles, presenting them to participants, and facilitating discussions on key findings, implications, and potential applications in precepting. These discussions encourage a deeper understanding of research literature and provide a platform for preceptors to exchange ideas, enhancing their collective knowledge and contributing to evidence-based precepting practices.

In summary, the combination of navigating respiratory care journals, database search strategies, critical reading skills, and discussions on selected articles empowers preceptors to actively engage with and apply the latest research in respiratory care education. These skills are crucial for staying informed, continuously improving precepting practices, and ensuring that educational approaches are evidence-based and relevant.

Chapter 9: Simulation-Based Skill Application

Objectives: 9

- Practical simulations of real-world precepting scenarios
- Application of learned concepts in simulated environments
- Feedback and improvement through simulation-based exercises

Simulation-Based Skill Application

Topic 1: Simulation 1 - Real-World Precepting Scenario 1

Objective: To immerse preceptors in a realistic simulation of a precepting scenario, allowing them to apply evidence-based practices and teaching strategies in a controlled environment.

Content:

1. *Introduction to Simulation:* Overview of the importance of simulation-based skill application in precepting.
2. *Scenario Description:* Presentation of a detailed real-world precepting scenario involving a respiratory care student.
3. *Application of Evidence-Based Practices:* Guidance on applying evidence-based practices, principles, and strategies in the simulated scenario.
4. *Debriefing:* Post-simulation discussion focusing on participants' decisions, challenges faced, and the alignment of actions with evidence-based precepting principles.

Simulation Script: Real-World Precepting Scenario

Objective:

To immerse preceptors in a realistic simulation of a precepting scenario, allowing them to apply evidence-based practices and teaching strategies in a controlled environment.

Introduction to Simulation: Overview of the Importance of Simulation-Based Skill Application in Precepting

Facilitator:

Welcome, everyone, to this simulation-based learning experience focused on real-world precepting scenarios. Today, we will explore the importance of simulation in enhancing your precepting skills. Simulations provide a safe environment to practice and refine teaching strategies, promoting experiential learning and decision-making. Let us dive into our first scenario.

Scenario Description:

Presentation of a Detailed Real-World Precepting Scenario Involving a Respiratory Care Student

Facilitator:

In this scenario, you are a respiratory care preceptor overseeing a student in a clinical setting. Your student is tasked with managing a complex patient case, and you will navigate various challenges commonly encountered in respiratory care education. Pay close attention to the details, as the scenario is designed to mirror real-world precepting experiences.

(Student enters the simulation room and begins interacting with the patient.)

Application of Evidence-Based Practices*:*

Guidance on Applying Evidence-Based Practices, Principles, and Strategies in the Simulated Scenario

Facilitator:

As you engage with the scenario, remember to apply evidence-based practices, principles, and teaching strategies. Consider the latest research findings and best practices in respiratory care education. Your decisions should align with evidence-informed precepting approaches. Feel free to adapt your teaching methods based on the evidence available.

(Preceptors observe the simulation, making decisions and offering guidance to the student.)

Debriefing:

Post-Simulation Discussion Focusing on Participants' Decisions, Challenges Faced, and the Alignment of Actions with Evidence-Based Precepting Principles

Facilitator:

Excellent effort, everyone. Now, let us gather for the debriefing session. Reflect on the decisions you made, the challenges faced, and how your actions align with evidence-based precepting principles. This is an opportunity for open discussion and learning from each other's experiences. Share your insights, and let's explore areas for improvement and reinforcement.

(Preceptors and the student engage in a facilitated discussion, reflecting on the simulation experience.)

This simulation script aims to create an immersive and educational experience for preceptors, allowing them to actively participate in a real-world precepting scenario. The script emphasizes the importance of evidence-based practices and principles, promoting a reflective learning environment during the debriefing session.

Topic 2: Simulation 2 - Real-World Precepting Scenario 2

Objective: To provide preceptors with another immersive simulation experience, reinforcing the application of evidence-based precepting skills in a different scenario.

Content:

1. *Overview of Simulation 2:* Introduction to the second precepting scenario, distinct from Simulation 1.

2. *Evidence-Based Decision-Making:* Emphasis on making evidence-based decisions in response to challenges presented in the simulation.
3. *Adaptation of Teaching Strategies:* Guidance on adapting teaching strategies based on the evolving needs and dynamics of the simulation.
4. *Debriefing and Reflection*: Facilitated discussion on the outcomes of Simulation 2, encouraging participants to reflect on their actions and identify areas for improvement.

Simulation Script: Introduction to the Second Precepting Scenario

Objective:

To introduce preceptors to the second precepting scenario, distinct from Simulation 1, emphasizing evidence-based decision-making, adaptation of teaching strategies, and facilitating a debriefing and reflection session.

Introduction: Overview of the Importance of the Second Simulation

Facilitator:

Welcome back, everyone. In our second simulation, we will explore a distinct precepting scenario, offering new challenges and opportunities for growth. This simulation builds on the principles discussed earlier, focusing on evidence-based decision-making and the adaptation of teaching strategies. Let us jump into the scenario.

Scenario Description: Presentation of the Second Precepting Scenario

Facilitator:

In this scenario, you will encounter a different set of challenges in a respiratory care education setting. The patient case and student interactions are designed to test your ability to make evidence-based decisions and adapt your teaching strategies to the evolving needs of the situation. Pay close attention to the unique dynamics presented in Simulation 2.

(Student enters the simulation room, and the scenario begins.)

Evidence-Based Decision-Making: Emphasis on Making Evidence-Based Decisions

Facilitator:

As you navigate this scenario, keep a focus on evidence-based decision-making. Consider the latest research findings, best practices, and guidelines in respiratory care education. Your decisions should be informed by evidence, ensuring that the actions you take align with the principles of evidence-based precepting.

(Preceptors observe and engage with the simulation, making decisions based on evidence.)

Adaptation of Teaching Strategies: Guidance on Adapting Teaching Strategies

Facilitator:

Adaptability is key in precepting. As the scenario unfolds, be prepared to adapt your teaching strategies based on the evolving needs and dynamics. Consider how you can tailor your approach to effectively support the student and address the challenges presented. Flexibility and responsiveness are valuable skills in precepting.

(Preceptors actively engage in adapting teaching strategies during the simulation.)

Debriefing and Reflection: Facilitated Discussion on the Outcomes of Simulation 2

Facilitator:

Excellent effort, everyone. Now, let us transition to the debriefing session. Reflect on the outcomes of Simulation 2, share your experiences, and discuss the decisions made. This is an opportunity for collaborative reflection, identifying areas of success and pinpointing areas for improvement. Let us engage in an open and insightful discussion.

(Preceptors and the student participate in a facilitated debriefing session.)

This simulation script aims to guide preceptors through the second precepting scenario, focusing on evidence-based decision-making, adaptation of teaching strategies, and fostering a reflective debriefing session. The script encourages active participation, discussion, and continuous learning within the context of respiratory care education.

Topic 3: Feedback Session - Improving Skills through Simulation

Objective: To create a structured feedback session where preceptors receive constructive feedback on their simulation performance, facilitating continuous improvement.

Content:

1. *Importance of Feedback in Learning:* Overview of the role of feedback in enhancing precepting skills.
2. *Simulation Debriefing Model:* Introduction to a debriefing model for structured feedback sessions.
3. *Peer Feedback Exchange:* Participants engage in providing constructive feedback to their peers based on the simulation experiences.
4. *Facilitated Group Reflection:* Facilitator-led group reflection on the feedback received, focusing on areas of strength and opportunities for enhancement.

Importance of Feedback in Learning: *Overview of the Role of Feedback in Enhancing Precepting Skills*

Feedback plays a crucial role in enhancing precepting skills within the context of respiratory care education. It serves as a powerful tool for professional development, providing preceptors with valuable insights into their teaching practices. By offering constructive feedback, preceptors can gain awareness of their strengths and areas for improvement, fostering a continuous learning mindset. The importance of feedback lies in its ability to guide preceptors toward evidence-based approaches, refine their decision-making, and promote effective communication in precepting scenarios.

Simulation Debriefing Model: *Introduction to a Debriefing Model for Structured Feedback Sessions*

A simulation debriefing model is an organized framework designed to structure feedback sessions after simulation-based learning experiences in respiratory care education. This model introduces a systematic approach to reflection and discussion, ensuring that the debriefing process is constructive and focused. It typically involves reviewing the simulation scenario, discussing participant decisions, and exploring the alignment of actions with evidence-based precepting principles. The debriefing model facilitates a comprehensive analysis of the simulation, allowing preceptors to extract meaningful insights and apply them to their future precepting practices.

Peer Feedback Exchange: *Participants Engage in Providing Constructive Feedback to Their Peers Based on the Simulation Experiences*

Peer feedback exchange involves participants actively engaging in providing constructive feedback to their peers based on their experiences in simulation-based precepting scenarios. This collaborative approach allows preceptors to share observations, insights, and recommendations with each other. Peer feedback fosters a supportive learning community where participants can benefit from diverse perspectives, gaining fresh insights into their precepting skills. It encourages a culture of mutual learning and growth, as preceptors contribute to each other's professional development through thoughtful and constructive feedback.

Facilitated Group Reflection: *Facilitator-Led Group Reflection on the Feedback Received, Focusing on Areas of Strength and Opportunities for Enhancement*

Facilitated group reflection is a structured process led by a facilitator to guide participants in reflecting on the feedback received during simulation-based learning experiences. The facilitator encourages participants to explore areas of strength in their precepting practices and identifies opportunities for enhancement. This group reflection session creates a collaborative environment where preceptors can collectively analyze their performance, share insights, and collectively brainstorm strategies for improvement. It enhances the impact of feedback by turning it into actionable insights that contribute to ongoing professional development in respiratory care education.

In summary, the importance of feedback in learning is paramount for preceptors in respiratory care education. Utilizing a simulation debriefing model, engaging in peer feedback exchange, and participating in facilitated group reflection collectively contribute to a holistic approach to enhancing precepting skills. These practices create a culture of continuous improvement, collaboration, and reflective learning among preceptors in the field of respiratory care.

CHAPTER 10: EMPOWERING PRACTITIONERS FOR FUTURE PRECEPTING ROLES

Objective:10

- Confidence-building exercises for preceptors
- Recognizing the role of preceptors in shaping the next generation
- Responsibilities and leadership in respiratory care education

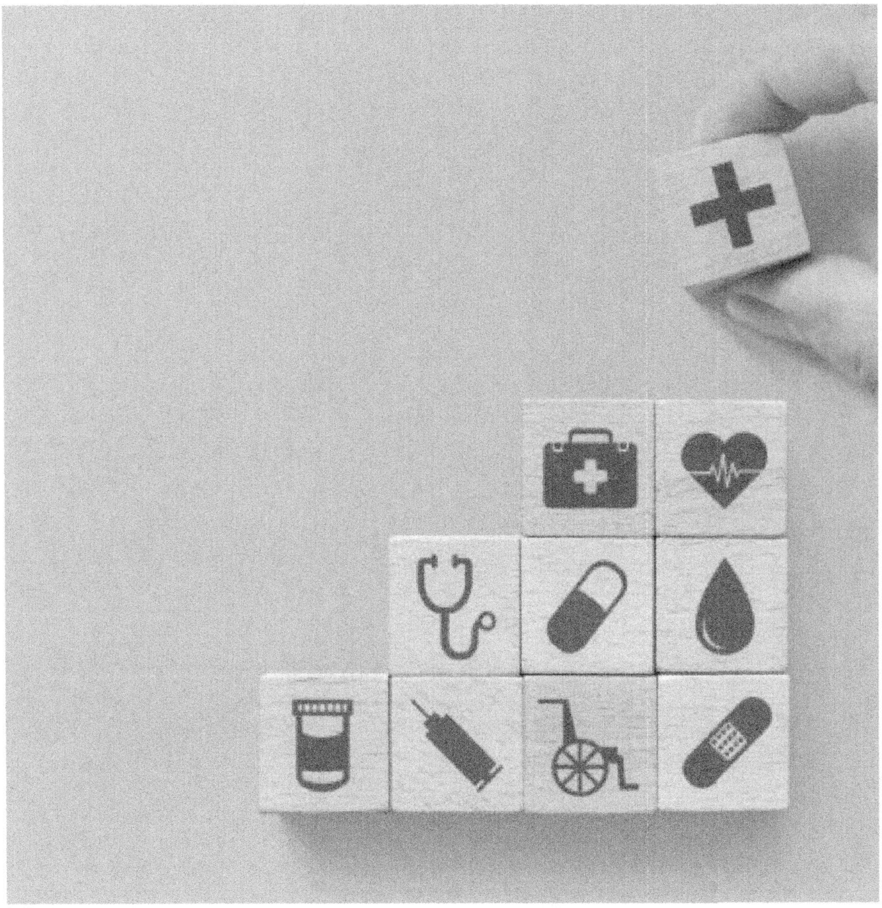

Confident and Empowered Preceptors

Topic 1: Confidence-Building Workshop: Precepting Exercises

Objective:

To provide preceptors with a confidence-building workshop incorporating precepting exercises that enhance their skills and empower them in their role.

Content:

1. **Introduction to the Workshop: The Significance of Confidence in Precepting**
   - Overview of the workshop's objective and the importance of confidence in effective precepting.
2. **Understanding and Addressing Precepting Challenges**
   - Discussion on familiar challenges faced by preceptors.
   - Strategies for building confidence in addressing challenges.
3. **Precepting Exercises: Practical Application**
   - Engaging preceptors in hands-on exercises related to real-world precepting scenarios.
   - Focus on decision-making, effective communication, and adapting teaching strategies.
4. **Group Reflection and Confidence Boosting**
   - Facilitated group reflection on the exercises.
   - Identification of areas where confidence has grown and strategies for continued improvement.

The Significance of Confidence in Precepting

Synopsis: Confidence in precepting is crucial for creating a positive learning environment and facilitating effective mentorship in respiratory care education. A confident preceptor inspires trust, fosters student engagement, and navigates challenges with resilience, contributing to enhanced learning experiences.

Points:

1. **Trust Building:**
   - Confidence instills trust in students, creating a supportive relationship that encourages open communication and shared learning.

2. **Role Modeling:**
    - Confident preceptors serve as role models, demonstrating competence and professionalism, which positively influences students' professional development.
3. **Effective Communication:**
    - Confidence enables clear and effective communication, facilitating the exchange of information, feedback, and guidance between preceptor and student.
4. **Adaptability:**
    - Confident preceptors are more adaptable in addressing unexpected challenges, providing a sense of stability and assurance to students in dynamic clinical settings.
5. **Motivation and Engagement:**
    - Confidence motivates students by instilling a sense of security and belief in their abilities, leading to increased engagement and participation in the learning process.

Understanding and Addressing Precepting Challenges

Synopsis: Recognizing and addressing precepting challenges is essential for maintaining a positive and effective preceptor-student dynamic in respiratory care education. Understanding these challenges allows preceptors to proactively implement strategies that enhance the overall learning experience.

Points:

1. **Identifying Common Challenges:**
    - Preceptors gain awareness of challenges such as communication barriers, varying student learning styles, and clinical uncertainties.
2. **Strategies for Building Resilience:**
    - Understanding challenges enables preceptors to develop resilience strategies, fostering an ability to adapt and overcome obstacles in the precepting process.
3. **Improving Communication Skills:**
    - Awareness of communication challenges prompts preceptors to hone their communication skills, ensuring clarity and effectiveness in conveying information to students.
4. **Tailoring Teaching Strategies:**

- Recognizing diverse learning styles allows preceptors to tailor teaching strategies, accommodating the individual needs of students, and optimizing the learning experience.

5. **Creating a Supportive Environment:**

- Proactively addressing challenges contributes to the creation of a supportive learning environment, enhancing the overall well-being of both preceptors and students.

Precepting Exercises: Practical Application

Synopsis: Precepting exercises with practical application immerse preceptors in real-world scenarios, allowing them to apply theoretical knowledge, refine their decision-making skills, and gain hands-on experience in a controlled environment.

Points:

1. **Firsthand Learning:**

- Practical exercises provide preceptors with hands-on learning opportunities, allowing them to actively engage in precepting scenarios.

2. **Decision-Making Skills:**

- Exercises focus on decision-making skills, enabling preceptors to practice making informed and evidence-based decisions in diverse clinical situations.

3. **Adapting Teaching Strategies:**

- Precepting exercises emphasize the adaptation of teaching strategies based on the evolving needs of the scenario, promoting flexibility and responsiveness.

4. **Real-World Scenario Simulation:**

- Simulation of real-world scenarios mirrors actual precepting experiences, preparing preceptors for the dynamic nature of clinical education.

5. **Integration of Evidence-Based Practices:**

- Practical application encourages the integration of evidence-based practices, aligning preceptors with current research and best practices in respiratory care education.

Group Reflection and Confidence Boosting

Synopsis: Group reflection sessions following precepting exercises provide a platform for preceptors to share experiences, receive constructive feedback, and collectively boost confidence. The supportive environment fosters a sense of community and empowers preceptors in their role.

Points:

1. **Sharing Experiences:**
   - Group reflection allows preceptors to share their experiences during precepting exercises, promoting open communication and collaborative learning.
2. **Constructive Feedback:**
   - Participants receive constructive feedback from peers and facilitators, offering valuable insights for improvement and reinforcing areas of strength.
3. **Identification of Growth Opportunities:**
   - Group reflection helps preceptors identify growth opportunities, encouraging a continuous learning mindset and promoting professional development.
4. **Collective Confidence Building:**
   - The collective sharing of experiences and feedback contributes to confidence building within the group, fostering a supportive community of preceptors.
5. **Facilitated Discussion:**
   - Facilitated discussions guide preceptors in exploring their decisions, challenges faced, and aligning actions with evidence-based precepting principles, enhancing overall confidence.

Topic 2: Recognizing the Role of Preceptors

Objective:

To deepen preceptors' understanding of their role in respiratory care education and empower them with a sense of purpose.

Content:

1. **Introduction to the Lecture: The Significance of the Preceptor Role**
   - Highlighting the pivotal role preceptors play in shaping the next generation of respiratory care practitioners.

2. **Responsibilities and Expectations of Preceptors**
   - Discussion on the core responsibilities of preceptors in guiding and mentoring students.
   - Exploration of expectations in various clinical and educational settings.
3. **Case Studies: Exemplary Preceptorship**
   - Presentation of case studies showcasing successful preceptor-student relationships.
   - Analysis of key attributes contributing to effective precepting.
4. **Interactive Discussion: Sharing Experiences and Insights**
   - Open forum for preceptors to share their experiences, challenges, and successes.
   - Facilitated discussion on strategies for overcoming common precepting obstacles.

The Significance of the Preceptor Role

Synopsis: The preceptor role is of paramount importance in respiratory care education, serving as a guiding force for students' professional development. Preceptors play a pivotal role in shaping future respiratory care practitioners by providing mentorship, fostering a positive learning environment, and exemplifying the standards of the profession.

Detailed Explanation:

1. **Guiding Professional Development:**
   - Preceptors guide students in the development of clinical and professional skills, ensuring they acquire the necessary competencies for success in respiratory care.
2. **Mentorship and Support:**
   - Preceptors serve as mentors, offering support, guidance, and encouragement throughout the learning journey. They play a crucial role in helping students navigate challenges and build confidence.
3. **Modeling Professionalism:**
   - The preceptor sets an example of professionalism, demonstrating ethical behavior, effective communication, and a commitment to evidence-based practice. This modeling shapes students' understanding of the standards and expectations in respiratory care.
4. **Creating a Positive Learning Environment:**

- The preceptor creates a positive and conducive learning environment, fostering open communication, trust, and collaboration. This environment enhances the overall learning experience for students.

5. **Ensuring Patient Safety:**

- Preceptors play a vital role in ensuring patient safety by overseeing students' clinical activities, providing supervision, and reinforcing adherence to safety protocols in respiratory care practices.

Responsibilities and Expectations of Preceptors

Synopsis: Preceptors in respiratory care education have multifaceted responsibilities that encompass guidance, supervision, and mentorship. Understanding these responsibilities and expectations is essential for fulfilling the preceptor role effectively.

Detailed Explanation:

1. **Clinical Supervision:**

- Preceptors are responsible for overseeing students' clinical activities, ensuring they apply theoretical knowledge in practical settings while upholding patient safety.

2. **Mentoring and Guidance:**

- Providing mentorship involves offering guidance, constructive feedback, and support to help students navigate challenges and develop both professionally and personally.

3. **Assessment and Evaluation:**

- Preceptors assess students' clinical competencies, providing objective feedback and conducting evaluations to ensure students meet established standards.

4. **Adapting Teaching Strategies:**

- Recognizing diverse learning styles, preceptors adapt teaching strategies to cater to individual needs, promoting an inclusive and effective learning experience.

5. **Communication with Educational Institutions:**

- Preceptors communicate with educational institutions, sharing insights on students' progress, contributing to the continuous improvement of the respiratory care curriculum.

Case Studies: Exemplary Preceptorship

Synopsis: Case studies highlighting exemplary preceptorship provide real-world examples of successful mentor-student relationships. These cases showcase effective communication, positive outcomes, and the impact of exceptional precepting on student development.

Detailed Explanation:

1. **Demonstrating Effective Precepting Practices:**
   - Case studies depict instances where preceptors demonstrated effective communication, adaptability, and leadership, contributing to positive learning experiences.
2. **Positive Student Outcomes:**
   - Exemplary preceptorship leads to positive student outcomes, including enhanced clinical competence, increased confidence, and a strong foundation for professional practice.
3. **Adapting to Varied Learning Styles:**
   - Successful case studies often involve preceptors who adapt their teaching strategies to accommodate diverse learning styles, ensuring that each student receives tailored support.
4. **Building Trust and Rapport:**
   - Case studies highlight the importance of building trust and rapport between preceptors and students, fostering a supportive and collaborative learning environment.
5. **Continuous Improvement:**
   - Exemplary precepting practices contribute to continuous improvement, both for preceptors and the educational institutions, reinforcing the importance of ongoing professional development.

Interactive Discussion: Sharing Experiences and Insights

Synopsis: Interactive discussions provide a platform for preceptors to share their experiences and insights, fostering a collaborative learning environment. The exchange of ideas and perspectives contributes to a rich pool of knowledge within the respiratory care education community.

Detailed Explanation:

1. **Open Forum for Experience Sharing:**
   - Interactive discussions create an open forum where preceptors can freely share their experiences, challenges, and successes, promoting transparency and mutual learning.
2. **Identification of Common Challenges:**

- Preceptors collaboratively identify common challenges, allowing for shared problem-solving and the exchange of strategies to overcome obstacles in the precepting process.

3. **Peer Learning and Support:**
    - Sharing experiences facilitates peer learning and support, as preceptors draw on collective insights to enhance their own practices and navigate complex scenarios.
4. **Professional Development Opportunities:**
    - Interactive discussions offer professional development opportunities by encouraging preceptors to learn from each other, stay informed about best practices, and explore innovative approaches to precepting.
5. **Building a Community of Support:**
    - Participating in interactive discussions contributes to the development of a community of support among preceptors, fostering a culture of collaboration, shared growth, and a collective commitment to excellence.

Topic 3: Group Activity: Responsibilities and Leadership in Respiratory Care Education

Objective:

To foster a sense of responsibility and leadership among preceptors in respiratory care education.

Content:

1. **Introduction to the Group Activity: Emphasizing Collective Leadership**
    - Overview of the importance of shared responsibility and leadership in the educational environment.
2. **Identifying Responsibilities in Respiratory Care Education**
    - Group activity to identify key responsibilities in precepting and respiratory care education.
    - Collaboration in developing a collective understanding of the preceptor's role.
3. **Leadership Skills Development: Strategies and Best Practices**
    - Presentation on effective leadership skills in the context of precepting.
    - Practical tips for cultivating leadership qualities.
4. **Group Discussion: Creating a Supportive Precepting Community**

- Facilitated discussion on building a community of support among preceptors.
- Sharing ideas for enhancing leadership and support networks.

Introduction to the Group Activity: Emphasizing Collective Leadership

Synopsis: The introduction to a group activity emphasizes collective leadership in respiratory care education. By fostering a sense of shared responsibility and collaboration, this approach aims to enhance the precepting community's effectiveness and create a supportive environment.

Detailed Explanation:

1. **Emphasis on Collaboration:**

    - The introduction highlights the importance of collaborative efforts within the precepting community, emphasizing that leadership is a collective responsibility rather than an individual endeavor.

2. **Shared Vision and Goals:**

    - Collective leadership encourages participants to align with a shared vision and common goals, promoting unity in achieving excellence in respiratory care education.

3. **Inclusivity and Participation:**

    - The introduction underscores the inclusivity of the group activity, encouraging active participation from all members to contribute their perspectives, experiences, and insights.

4. **Building a Community:**

    - The activity aims to build a sense of community among preceptors, emphasizing that each member plays a vital role in the success of the overall educational experience.

Identifying Responsibilities in Respiratory Care Education

Synopsis: Identifying responsibilities in respiratory care education is essential for preceptors to understand their roles and contribute effectively to the learning environment. This step involves recognizing the multifaceted aspects of precepting and their impact on student development.

Detailed Explanation:

1. **Clinical Supervision and Guidance:**

    - Participants identify their responsibility for clinical supervision, ensuring students receive guidance and support in applying theoretical knowledge to real-world scenarios.

2. **Mentoring and Support:**

- Preceptors recognize their role as mentors, providing ongoing support, constructive feedback, and guidance to facilitate the professional development of respiratory care students.

3. **Assessment and Evaluation:**
   - Participants acknowledge their responsibility for assessing students' clinical competencies and conducting evaluations, ensuring students meet the required standards for respiratory care practice.
4. **Adapting Teaching Strategies:**
   - Identifying responsibilities includes recognizing the need to adapt teaching strategies to accommodate diverse learning styles, creating an inclusive and effective learning experience.
5. **Communication with Educational Institutions:**
   - Preceptors understand their responsibility to communicate effectively with educational institutions, sharing insights on students' progress to contribute to continuous improvement.

Leadership Skills Development: Strategies and Best Practices

Synopsis: Leadership skills development focuses on equipping preceptors with effective strategies and best practices to enhance their leadership capabilities in respiratory care education. This step aims to empower preceptors to lead with confidence and impact.

Detailed Explanation:

1. **Effective Communication:**
   - Leadership skills development emphasizes the importance of effective communication, enabling preceptors to convey information clearly, foster understanding, and promote collaboration.
2. **Adaptability and Flexibility:**
   - Strategies include fostering adaptability and flexibility, enabling preceptors to navigate dynamic clinical environments and respond to evolving challenges with resilience.
3. **Motivational Techniques:**
   - Preceptors learn motivational techniques to inspire and engage students, fostering a positive learning environment that encourages active participation and enthusiasm.

4. **Problem-Solving Abilities:**
    - Leadership skills development includes enhancing problem-solving abilities, equipping preceptors to address challenges, make informed decisions, and contribute to continuous improvement.
5. **Collaborative Decision-Making:**
    - Participants learn strategies for collaborative decision-making, promoting inclusivity and involving the precepting community in key decisions related to respiratory care education.

Group Discussion: Creating a Supportive Precepting Community

Synopsis: Group discussions aim to create a supportive precepting community by providing a platform for open dialogue, idea exchange, and mutual support. This collaborative environment fosters a culture of shared learning and continuous improvement.

Detailed Explanation:

1. **Open Forum for Communication:**
    - Group discussions serve as an open forum for preceptors to communicate openly, share experiences, and express insights, fostering transparency and trust.
2. **Peer Learning and Support:**
    - Participants engage in peer learning and support, drawing on collective insights to enhance their own practices, navigate challenges, and contribute to the professional growth of their peers.
3. **Professional Development Opportunities:**
    - Group discussions offer opportunities for professional development by encouraging preceptors to stay informed about best practices, exchange innovative ideas, and explore innovative approaches to precepting.
4. **Building a Community of Support:**
    - The discussion focuses on building a community of support among preceptors, where shared experiences and collaborative efforts contribute to a supportive and thriving learning environment.
5. **Promoting Continuous Improvement:**

- Group discussions contribute to continuous improvement by encouraging participants to reflect on their experiences, share lessons learned, and collectively identify strategies for enhancing precepting practices.

REFERENCES

AARC Clinical Practice Guidelines. (2020). Interpretation of Arterial Blood Gases. American Association for Respiratory Care.

Accreditation Standards and Guidelines for the Professional Program in Respiratory Care. (2018). Commission on Accreditation for Respiratory Care (CoARC).

Alinier, G., Hunt, B., Gordon, R., & Harwood, C. (2006). Effectiveness of intermediate-fidelity simulation training technology in undergraduate nursing education. Journal of Advanced Nursing, 54(3), 359-369.

American Association for Respiratory Care (AARC) Clinical Practice Guidelines. Retrieved from https://www.aarc.org

American Association for Respiratory Care (AARC). (2021). *AARC Code of Ethics.*

American Association for Respiratory Care (AARC). (2021). AARC Statement on Continuing Education.

American Association for Respiratory Care (AARC). (2022). Evidence-Based Clinical Practice Guidelines. https://www.aarc.org/resources/clinical-resources/evidence-based-guidelines/

American Association for Respiratory Care (AARC). (2022). https://www.aarc.org/

American Association for Respiratory Care (AARC). (2022). https://www.aarc.org/

American Association for Respiratory Care. (2019). AARC Statement of Ethics and Professional Conduct. https://www.aarc.org/wp-content/uploads/2019/09/Statement-of-Ethics.pdf

American Association for Respiratory Care. (2019). Code of Ethics. https://www.aarc.org/resources/ethics-and-standards/code-of-ethics/

American Association for Respiratory Care. (2020). Clinical Practice Guidelines. https://www.aarc.org/resources/clinical-resources/clinical-practice-guidelines/

American Association for Respiratory Care. (2021). *AARC U.* Retrieved from https://www.aarc.org/education/aarc-university/

American Association for Respiratory Care. (2021). Clinical Practice Guidelines. https://www.aarc.org/resources/clinical-resources/clinical-practice-guidelines/

American Association for Respiratory Care. (2021). *Education.* Retrieved from https://www.aarc.org/education/

American Association for Respiratory Care. (2021). Respiratory Care in Health Systems. https://www.aarc.org/resources/clinical-resources/health-systems/

American Association for Respiratory Care. (2021). Respiratory Care Licensure. https://www.aarc.org/resources/licensure/

American Association for Respiratory Care. (n.d.). Resources for Respiratory Therapists. https://www.aarc.org/resources/

American Health Information Management Association. (2021). Legal Health Record Standards. https://www.ahima.org/

American Heart Association. (2015). *Advanced Cardiovascular Life Support (ACLS) Provider Manual.* Dallas, TX: American Heart Association.

American Heart Association. (2015). *Basic life support (BLS) provider manual.* Dallas, TX: American Heart Association.

American Lung Association. (2021). Patient Rights. https://www.lung.org/

American Lung Association. (2021). Respiratory Therapist Licensure Requirements. https://www.lung.org/

American Medical Association. (2019). Principles of Medical Ethics. https://www.ama-assn.org/delivering-care/ethics/principles-medical-ethics

American Psychological Association. (2020). Publication Manual of the American Psychological Association (7th ed.).

American Psychological Association. (2022). *Publication Manual of the American Psychological Association* (7th ed.). APA.

American Thoracic Society. (2017). Pulmonary Function Testing. https://www.thoracic.org/patients/patient-resources/resources/pulmonary-function-testing.pdf

Anderson, L. K., & Thompson, M. J. (Year). "Code Blue Teams: Enhancing Emergency Response in Healthcare." Publisher.

Apfelbaum, J. L., Hagberg, C. A., Caplan, R. A., Blitt, C. D., Connis, R. T., Nickinovich, D. G., ... & American Society of Anesthesiologists Committee on Standards and Practice Parameters. (2013). *Practice guidelines for management of the difficult airway: an updated report by the American Society of Anesthesiologists Task Force on Management of the Difficult Airway.* Anesthesiology, 118(2), 251-270.

Association for the Advancement of Medical Instrumentation. (2019). Comprehensive Guide to Steam Sterilization and Sterility Assurance in Health Care Facilities. ANSI/AAMI ST79:2017. Arlington, VA: AAMI.

Baltierra, N. B., Choudhury, A., & Duggan, C. (2017). Overview of the Role of Nutrition in Chronic Obstructive Pulmonary Disease. Current Nutrition Reports, 6(4), 266–271. https://doi.org/10.1007/s13668-017-0210-y

Barnes, P. J. (2010). Inhaled corticosteroids. Pharmaceuticals, 3(3), 514-540.

Bass, B. M., & Riggio, R. E. (2006). Transformational leadership (2nd ed.). Psychology Press.

Beauchamp, T. L., & Childress, J. F. (2019). Principles of Biomedical Ethics. Oxford University Press.

Beauchamp, T. L., & Childress, J. F. (2019). Principles of Biomedical Ethics (8th ed.). Oxford University Press.

Beauchamp, T. L., & Childress, J. F. (2019). *Principles of biomedical ethics.* Oxford University Press.

Beauchamp, T. L., & Childress, J. F. (2019). Principles of biomedical ethics. Oxford University Press.

Beauchamp, T. L., & Childress, J. F. (2019). Principles of Biomedical Ethics (8th ed.). Oxford University Press.

Beckett, R. D. (2008). *Patient-centered communication during the respiratory care interview.* Respiratory Care, 53(6), 722-733.

Beers, M. H., & Berkow, R. (Eds.). (2020). The Merck Manual of Diagnosis and Therapy. Merck Research Laboratories.

Benner, P., Sutphen, M., Leonard, V., & Day, L. (2009). *Educating nurses: A call for radical transformation.* John Wiley & Sons.

Berkman, N. D., Sheridan, S. L., Donahue, K. E., Halpern, D. J., & Crotty, K. (2011). Low health literacy and health outcomes: An updated systematic review. Annals of Internal Medicine, 155(2), 97–107. https://doi.org/10.7326/0003-4819-155-2-201107190-00005

Betancourt, J. R., Green, A. R., & Carrillo, J. E. (2002). Cultural Competence in Health Care: Emerging Frameworks and Practical Approaches. The Commonwealth Fund. https://www.commonwealthfund.org/sites/default/files/documents/___media_files_publications_fund_report_2002_oct_cultural_competence_emerging_frameworks_carrillo_culturalcompethcare_576_pdf.pdf

Betancourt, J. R., Green, A. R., & Carrillo, J. E. (2002). Cultural competence in health care: Emerging frameworks and practical approaches. *The Commonwealth Fund.*

Betancourt, J. R., Green, A. R., Carrillo, J. E., & Ananeh-Firempong, O. (2003). *Defining cultural competence: a practical framework for addressing racial/ethnic disparities in health and health care.* Public Health Reports, 118(4), 293-302.

Bing-You, R., Hayes, V., Varaklis, K., & Trowbridge, R. (2016). Feedback for learners in medical education: What is known? *The Clinical Teacher, 13*(1), 9–14.

Brathwaite, A. C., & Majumdar, B. (2006). Cultural competence in medical education: Demographic differences associated with medical student communication styles and clinical clerkship feedback. *Patient Education and Counseling, 64*(1-3), 125-132.

Brown, C. E., Wickline, M. A., Ecoff, L., & Glaser, D. (2009). Nursing practice, knowledge, attitudes and perceived barriers to evidence-based practice at an academic medical center. *Journal of Advanced Nursing, 65*(2), 371-381.

Buckley, H., Hensman, M., Thomas, S., & Wetherell, M. (2019). The development of preceptor self-efficacy through the use of guided reflective journaling. *Nurse Education Today, 76*, 52-57.

Burns, J. M. (1978). Leadership. Harper & Row.

Campinha-Bacote, J. (2002). The process of cultural competence in the delivery of healthcare services: A model of care. *Journal of Transcultural Nursing, 13*(3), 181-184.

Cazzola, M., & Matera, M. G. (2008). Bronchodilators: current and future. Clinics in Chest Medicine, 29(1), 45-62.

Centers for Disease Control and Prevention. (2021). Medication Safety Program. https://www.cdc.gov/medicationsafety/index.html

Centers for Disease Control and Prevention. (2022). Guideline for Disinfection and Sterilization in Healthcare Facilities. https://www.cdc.gov/infectioncontrol/guidelines/disinfection/index.html

Centers for Disease Control and Prevention. (2022). Infection Control Basics. https://www.cdc.gov/infectioncontrol/basics/index.html

Centers for Disease Control and Prevention. (2022). Infection Control in Healthcare Settings. https://www.cdc.gov/infectioncontrol/index.html

Centers for Disease Control and Prevention. (2022). Isolation Precautions. https://www.cdc.gov/infectioncontrol/guidelines/isolation/index.html

Centers for Medicare & Medicaid Services. (2021). Electronic Health Records (EHR) Incentive Programs. https://www.cms.gov/Regulations-and-Guidance/Legislation/EHRIncentivePrograms

Centers for Medicare & Medicaid Services. (2021). Regulations and Guidance. https://www.cms.gov/Regulations-and-Guidance

Centers for Medicare & Medicaid Services. (n.d.). https://www.cms.gov/

Certification Board for Infection Control and Epidemiology (CBIC). (2022). https://www.cbic.org/

Chatburn, R. L. (2017). Fundamentals of Mechanical Ventilation: A Short Course on the Theory and Application of Mechanical Ventilators. CreateSpace Independent Publishing Platform.

Chua, C., & Hegney, D. (2017). Mentoring within the nursing profession in Australia: A review. *Collegian, 24*(5), 521–527.

Clutter, J., & Mayo, A. M. (2018). Mentoring in respiratory care. *Respiratory Care Education Annual, 27*(1), 37–42.

Coiera, E. (2017). Communication systems in healthcare. *Clinical Biochemist Reviews, 38*(4), 197–200.

Collins, P. F., Yang, I. A., Chang, Y. C., Vaughan, A., & McDonald, C. F. (2017). Respiratory health and inhalation exposure to particulate heavy metals in the outdoor air among community residents in Australia. Atmospheric Environment, 156, 146–153. https://doi.org/10.1016/j.atmosenv.2017.02.032

Commission on Accreditation for Respiratory Care (CoARC). (2020). Accreditation Standards for Respiratory Care Education.

Cook, T. M., & MacDougall-Davis, S. R. (2011). *Complications and failure of airway management.* British Journal of Anaesthesia, 109(suppl_1), i68-i85.

Coursera. (2022). https://www.coursera.org/

Cruess, R. L., Cruess, S. R., & Steinert, Y. (2016). Teaching Medical Professionalism: Supporting the Development of a Professional Identity. Cambridge University Press

Dargin, J., & Emlet, L. (2015). *Respiratory Care Procedures in the Home.* Respiratory Care, 60(6), 851-866.

Davis, A. H., & Schonfeld, T. L. (2020). Investing in Professional Development: The Why, What, and How of Continuing Education for Early Care and Education Professionals. Springer.

Davis, D., Bordage, G., Moores, L. K., & Bennett, N. (2001). The science of writing: Theories, methods, individual differences, and applications. *Academic Medicine, 76*(10), S85-S87.

Dieckmann, P., Gaba, D., Rall, M., & Topping, M. (2007). The WICS approach to debriefing: A conceptual framework based on review of literature and analysis of debriefing practices in simulation. Simulation in Healthcare, 2(3), 183–190. https://doi.org/10.1097/SIH.0b013e3180315539

Dieckmann, P., Molin Friis, S., Lippert, A., Østergaard, D., & Nakstad, A. R. (2012). *The art and science of simulation in healthcare.* Simulation & Gaming, 43(3), 297-318.

Downing, S. M. (2003). Validity: On the meaningful interpretation of assessment data. *Medical Education, 37*(9), 830–837.

Edmonson, A. C. (2019). *The fearless organization: Creating psychological safety in the workplace for learning, innovation, and growth.* John Wiley & Sons.

Effing, T. W., Bourbeau, J., Vercoulen, J., Apter, A. J., Coultas, D., Meek, P., van der Valk, P., Partridge, M. R., Palen, J., Rabe, K. F., & van der Palen, J. (2012). Self-management programmes for COPD: Moving forward. Chronic Respiratory Disease, 9(1), 27–35. https://doi.org/10.1177/1479972311430338

Epstein, R. M., & Hundert, E. M. (2002). Defining and assessing professional competence. *JAMA, 287*(2), 226-235.

Epstein, R. M., & Street Jr, R. L. (2007). *Patient-Centered Communication in Cancer Care: Promoting Healing and Reducing Suffering.* National Cancer Institute, NIH Publication No. 07-6225.

Fanning, R. M., & Gaba, D. M. (2007). *The role of debriefing in simulation-based learning.* Simulation in Healthcare, 2(2), 115-125.

Fleming, N. D., & Mills, C. (1992). Not another inventory, rather a catalyst for reflection. *To improve the academy, 11*(1), 137-155.

Flott, E. A., & Linden, L. (2016). The clinical learning environment in nursing education: A concept analysis. *Journal of Advanced Nursing, 72*(3), 501-513.

Frerk, C., Mitchell, V. S., McNarry, A. F., Mendonca, C., Bhagrath, R., Patel, A., ... & Pandit, J. J. (2015). *Difficult Airway Society 2015 guidelines for management of unanticipated difficult intubation in adults.* British Journal of Anaesthesia, 115(6), 827-848.

Frosch, D. L., May, S. G., Rendle, K. A. S., Tietbohl, C., & Elwyn, G. (2012). Authoritarian physicians and patients' fear of being labeled 'difficult' among key obstacles to shared decision making. Health Affairs, 31(5), 1030–1038. https://doi.org/10.1377/hlthaff.2011.0576

Gauthier, A., Cherney, A., & Stewart, T. (2015). Competency assessment in simulation-based education: A review of the literature. *Simulation in Healthcare, 10*(1), 27–34.

Global Initiative for Asthma (GINA). (2022). Global Strategy for Asthma Management and Prevention. https://ginasthma.org/gina-reports/

Global Initiative for Asthma. (2021). Global Strategy for Asthma Management and Prevention. https://ginasthma.org/wp-content/uploads/2021/05/GINA-Main-Report-2021-V2-WMS.pdf

Global Initiative for Chronic Obstructive Lung Disease (GOLD). (2021). Global strategy for the diagnosis, management, and prevention of chronic obstructive pulmonary disease.

Global Initiative for Chronic Obstructive Lung Disease. (2021). Global Strategy for the Diagnosis, Management, and Prevention of Chronic Obstructive Pulmonary Disease. https://goldcopd.org/wp-content/uploads/2020/11/GOLD-REPORT-2021-v1.1-25Nov20_WMV.pdf

Goldstein, R. S., & Ghezzo, H. (2001). *Optimal drug therapy in patients with COPD—new perspectives.* COPD: Journal of Chronic Obstructive Pulmonary Disease, 14(2), 223-232.

Gomersall, C. D., & Joynt, G. M. (2016). ABG Interpretation Made Easy. Elsevier.

Greenhalgh, T. (2014). *How to read a paper: The basics of evidence-based medicine.* John Wiley & Sons.

Greenleaf, R. K. (2008). The servant as leader. Robert K. Greenleaf Center.

Harris, M. R. (2015). Ethics in health services management. Routledge.

Hattie, J., & Timperley, H. (2007). The power of feedback. *Review of Educational Research, 77*(1), 81–112.

Health-Related Quality-of-Life Domains and Daily Step Counts in COPD: A Randomized Controlled Trial. Chest, 148(1), 128–137. https://doi.org/10.1378/chest.14-2388

Hess, D. (2007). *Chest physical therapy in the intensive care unit.* Respiratory Care, 52(7), 810-818.

Hess, D. R. (2019). *Respiratory care year in review 2018: airway management.* Respiratory Care, 64(5), 597-605.

Hill, J., & Watson, R. (2016). Creating an effective and supportive learning environment for nursing students: The importance of preceptorship. *Nurse Education in Practice, 16*(1), 67-71.

Holzhausen, Y., Maaz, A., Renz, A., Bosch, J., Peters, H., & Mitzkat, A. (2017). Competency-based undergraduate medical education: A literature review and implementation recommendations. *GMS Journal for Medical Education, 34*(3), Doc29.

Huber, D. (2017). *Leadership and nursing care management.* Elsevier Health Sciences.

Institute for Healthcare Improvement. (2022). Science of Improvement: How to Improve. https://www.ihi.org/resources/Pages/HowtoImprove/default.aspx

Issenberg, S. B., McGaghie, W. C., Petrusa, E. R., Lee Gordon, D., & Scalese, R. J. (2005). Features and uses of high-fidelity medical simulations that lead to effective learning: A BEME systematic review. Medical Teacher, 27(1), 10–28. https://doi.org/10.1080/01421590500046924

Issenberg, S. B., McGaghie, W. C., Petrusa, E. R., Lee Gordon, D., & Scalese, R. J. (2005). *Features and uses of high-fidelity medical simulations that lead to effective learning: a BEME systematic review.* Medical Teacher, 27(1), 10-28.

Issenberg, S. B., McGaghie, W. C., Petrusa, E. R., Lee Gordon, D., & Scalese, R. J. (2005). Features and uses of high-fidelity medical simulations that lead to effective learning: A BEME systematic review. *Medical Teacher, 27*(1), 10–28.

Jaber, S., Jung, B., & Chanques, G. (2010). *Endotracheal intubation and extubation in the ICU.* Critical Care Clinics, 26(3), 279-289.

Jeffries, P. R. (2005). A framework for designing, implementing, and evaluating simulations used as teaching strategies in nursing. Nursing Education Perspectives, 26(2), 96-103.

Jeffries, P. R. (2012). Simulation in Nursing Education: From Conceptualization to Evaluation. National League for Nursing.

Jeffries, P. R. (2012). *Simulation in nursing education: From conceptualization to evaluation.* National League for Nursing.

Joint Commission. (2021). Sentinel Event Data - Root Causes by Event Type. https://www.jointcommission.org/-/media/tjc/documents/standards/national-patient-safety-goals/2021/sed_root_causes_2021.pdf

Jolly, B. (2015). *Peer and self-assessment in the health professions: A user's guide.* AMEE Guide No. 81. Medical Teacher, 37(3), 252-265.

Jonsen, A. R., Siegler, M., & Winslade, W. J. (2015). Clinical Ethics: A Practical Approach to Ethical Decisions in Clinical Medicine (8th ed.). McGraw-Hill Education.

Juckett, G. (2013). Cross-Cultural Medicine. American Family Physician, 88(4), 248–254. https://www.aafp.org/afp/2013/0815/p248.html

Kacmarek, R. M., Stoller, J. K., & Heuer, A. J. (2016). Egan's Fundamentals of Respiratory Care. Elsevier Health Sciences.

Kacmarek, R. M., Stoller, J. K., & Heuer, A. J. (2016). Egan's Fundamentals of Respiratory Care. Elsevier.

Kardong-Edgren, S., Adamson, K. A., & Fitzgerald, C. (2010). A review of currently published evaluation instruments for human patient simulation. Clinical Simulation in Nursing, 6(1), e25-e35.

Kelly, P., & Hager, P. (2018). Enhancing the quality of clinical placement and preceptorship in final placements for undergraduate paramedicine students: A qualitative study. *Nurse Education Today, 60*, 56-61.

Khan Academy. (2022). https://www.khanacademy.org/

Knowles, M. S. (1984). Andragogy in action: Applying modern principles of adult education. Jossey-Bass.

Knowles, M. S., Holton III, E. F., & Swanson, R. A. (2015). *The adult learner: The definitive classic in adult education and human resource development.* Routledge.

Kogan, J. R., Holmboe, E. S., & Hauer, K. E. (2009). Tools for direct observation and assessment of clinical skills of medical trainees: A systematic review. *Journal of the American Medical Association, 302*(12), 1316–1326.

Kolb, D. A. (2014). *Experiential learning: Experience as the source of learning and development.* FT press.

Kumas-Tan, Z., & Beagan, B. (2015). Diversity and Cultural Competence Training in Health Care Organizations: Hallmarks of Success. Journal of Health Organization and Management, 29(6), 777–796. https://doi.org/10.1108/JHOM-06-2013-0126

Laube, B. L., Janssens, H. M., de Jongh, F. H., Devadason, S. G., Dhand, R., Diot, P., Everard, M. L., Horvath, I., Navalesi, P., Voshaar, T., & Chrystyn, H. (2011). What the pulmonary specialist should know about the new inhalation therapies. European Respiratory Journal, 37(6), 1308–1331. https://doi.org/10.1183/09031936.00166410

Lavorini, F., Corrigan, C., Barnes, P., Dekhuijzen, R., Levy, M., Pedersen, S., & Hillyer, E. V. (2010). Retail sales of inhalation devices in European countries: So much for a global policy. Respiratory Medicine, 104(2), 191–195. https://doi.org/10.1016/j.rmed.2009.07.017

Leonard, M., Graham, S., & Bonacum, D. (2004). *The human factor: the critical importance of effective teamwork and communication in providing safe care.* Quality and Safety in Health Care, 13(Suppl 1), i85-i90.

Leonard, M., Graham, S., & Bonacum, D. (2004). The human factor: The critical importance of effective teamwork and communication in providing safe care. *Quality & Safety in Health Care, 13*(Suppl 1), i85–i90.

Levett-Jones, T., & Bourgeois, S. (2014). The clinical placement: An essential guide for nursing students. *Elsevier Health Sciences.*

Levy, M. L., & Carroll, W. D. (2019). Pediatric Respiratory Medicine. Elsevier.

Lie, D., & Lee-Rey, E. (2013). Gender and the clinical encounter: A study of academic trainees. *Journal of General Internal Medicine, 28*(8), 1097-1103.

LinkedIn. (2022). https://www.linkedin.com/

Lioce, L., Meakim, C. H., Fey, M. K., Chmil, J. V., Mariani, B., Alinier, G., ... & Waxman, K. (2015). Standards of best practice: Simulation standard IX: Simulation design. *Clinical Simulation in Nursing, 11*(6), 309-315.

Lo, B., & Rubenfeld, G. D. (2005). *Palliative sedation in dying patients: "we turn to it when everything else hasn't worked".* JAMA, 294(14), 1810-1816.

MacIntyre, N. R., & Hess, D. R. (2020). Ventilator Management Strategies for Critical Care. McGraw-Hill Education.

Mann, K., Gordon, J., & MacLeod, A. (2009). *Reflection and reflective practice in health professions education: a systematic review.* Advances in Health Sciences Education, 14(4), 595-621.

Marquis, B. L., & Huston, C. J. (2017). *Leadership roles and management functions in nursing: Theory and application.* Wolters Kluwer.

McCarthy, B., Casey, D., Devane, D., Murphy, K., Murphy, E., & Lacasse, Y. (2015). Pulmonary rehabilitation for chronic obstructive pulmonary disease. Cochrane Database of Systematic Reviews, 2, CD003793. https://doi.org/10.1002/14651858.CD003793.pub3

Melnyk, B. M., & Fineout-Overholt, E. (Eds.). (2019). *Evidence-based practice in nursing & healthcare: A guide to best practice.* Wolters Kluwer.

Merriam, S. B., & Bierema, L. L. (2013). Adult learning: Linking theory and practice. Jossey-Bass.

Morton, P. G., & Fontaine, D. K. (2018). Critical Thinking and Clinical Reasoning in the Health Sciences: An International Multidisciplinary Teaching Anthology. Elsevier Health Sciences.

Moy, M. L., Collins, R. J., Martinez, C. H., Kadri, R., Roman, P., Holleman, R. G., Kim, H. M., Nguyen, H. Q., Cohen, M. D., & Goodrich, D. E. (2015). An Internet-Mediated Pedometer-Based Program Improves

National Academies of Sciences, Engineering, and Medicine. (2016). Health Literacy and Communication Strategies in Oncology: Proceedings of a Workshop. National Academies Press. https://www.ncbi.nlm.nih.gov/books/NBK436845/

National Board for Respiratory Care (NBRC). (2021). Clinical Simulation Examination Policies and Procedures Manual.

National Board for Respiratory Care (NBRC). (2022). Continuing Competency Program Guide.

National Board for Respiratory Care (NBRC). (2022). Policy and Procedure Manual.

National Heart, Lung, and Blood Institute (NHLBI). (2021). Asthma Care Quick Reference: Diagnosing and Managing Asthma.

Nolan, J. P., Soar, J., Zideman, D. A., Biarent, D., Bossaert, L. L., Deakin, C., ... & European Resuscitation Council. (2010). *European Resuscitation Council Guidelines for Resuscitation 2010 Section 1. Executive summary.* Resuscitation, 81(10), 1219-1276.

Northouse, P. G. (2018). Leadership: Theory and practice (8th ed.). Sage Publications.

Northouse, P. G. (2018). *Leadership: Theory and Practice.* Sage Publications.

Olson, J., & Hollen, P. (2019). *Calculate with Confidence.* Elsevier.

Pashler, H., McDaniel, M., Rohrer, D., & Bjork, R. (2009). Learning styles: Concepts and evidence. *Psychological Science in the Public Interest, 9*(3), 105-119.

Pendleton, D., Schofield, T., Tate, P., & Havelock, P. (1984). *The consultation: An approach to learning and teaching.* Oxford University Press.

Pérez-Stable, E. J. (2019). Language access and the quality of healthcare: What do we know? Findings and implications for the California SORH. California Journal of Health Promotion, 7(1), 4–13. https://doi.org/10.32398/cjhp.v7i1.1940

Pryor, J. A., & Prasad, S. A. (2008). *Physiotherapy for respiratory and cardiac problems.* Elsevier Health Sciences.

Publication Manual of the American Psychological Association: The Official Guide to APA Style. (2019). United States: American Psychological Association.

PubMed. (2022). https://pubmed.ncbi.nlm.nih.gov/

Rall, M., Gaba, D., & Flin, R. (2008). *Human performance and patient safety.* In Patient Safety (pp. 155-163). Springer.

Ramani, S., & Krackov, S. K. (2012). Twelve tips for giving feedback effectively in the clinical environment. *Medical Teacher, 34*(10), 787–791.

Rankin, J. A., Stallard, B., & Ritcey, B. (2017). *Managing anxiety in the emergency department: a review of psychotropic oral anxiolytics.* The Journal of Emergency Medicine, 52(4), 458-466.

Rau, J. L. (2001). *The beta-adrenergic receptors: New directions in bronchodilator therapy.* Chest, 119(6), 1820-1836.

Respiratory Therapy Forum. (n.d.). https://www.rtforum.com/

Rest, J. R. (1984). Moral development: Advances in research and theory. Praeger.

Rudolph, J. W., Simon, R., Rivard, P., Dufresne, R. L., & Raemer, D. B. (2007). Debriefing with good judgment: Combining rigorous feedback with genuine inquiry. *Anesthesiology clinics, 25*(2), 361-376.

Sackett, D. L., Rosenberg, W. M., Gray, J. A., Haynes, R. B., & Richardson, W. S. (1996). Evidence based medicine: what it is and what it isn't. *BMJ, 312*(7023), 71-72.

Salas, E., Sims, D. E., & Burke, C. S. (2005). *Is there a "big five" in teamwork?* Small Group Research, 36(5), 555-599.

Sandars, J. (2009). *The use of reflection in medical education: AMEE Guide No. 44.* Medical Teacher, 31(8), 685-695.

Sargeant, J., Mann, K., Sinclair, D., Van der Vleuten, C., & Metsemakers, J. (2008). Challenges in multisource feedback: Intended and unintended outcomes. *Medical Education, 42*(6), 528–534.

Savery, J. R., & Duffy, T. M. (1996). Problem-based learning: An instructional model and its constructivist framework. Educational Technology, 36(5), 31-38.

Sawatzky, J.-A. V., Beresford, B., & Stelmaschuk, S. (2012). Enhancing nurses' empowerment: The role of paraprofessional support. Journal of Advanced Nursing, 68(7), 1598–1608. https://doi.org/10.1111/j.1365-2648.2011.05840.x

Sawyer, T., Eppich, W., Brett-Fleegler, M., Grant, V., Cheng, A., & Team, I. D. E. A. L. S. (2016). *More Than One Way to Debrief: A Critical Review of Healthcare Simulation Debriefing Methods.* Simulation in Healthcare, 11(3), 209-217.

Schillinger, D., Piette, J., Grumbach, K., Wang, F., Wilson, C., Daher, C., Leong-Grotz, K., Castro, C., Bindman, A. B., & Bibbins-Domingo, K. (2003). Closing the loop: Physician communication with diabetic patients who have low health literacy. Archives of Internal Medicine, 163(1), 83–90. https://doi.org/10.1001/archinte.163.1.83

Schols, A. M. W. J., & Ferreira, I. M. (2009). Nutritional assessment and therapy in COPD: A European Respiratory Society statement. European Respiratory Journal, 33(6), 1165–1182. https://doi.org/10.1183/09031936.00080308

Slowther, A., Johnston, C., Goodall, J., Hope, T., & Bestwick, J. (2004). *Development of clinical ethics services in the UK: a national survey.* Journal of Medical Ethics, 30(6), 555-561.

Smith, A.B., & Jones, C.D. (2020). Early Mobility in the Ventilated Patient: A Review. Respiratory Care, 65(8), 1185-1195. https://doi.org/10.4187/respcare.07938

Smith, P. (Ed.). (2019). Professional development in education: New paradigms and practices. Springer.

Society for Simulation in Healthcare. (2019). Standards for Simulation-Based Education: Enhancing Learner Outcomes in Health Professions Education. Lippincott Williams & Wilkins.

Stone, J. H. (2006). Academic clinical training in respiratory care: Meeting the cultural diversity challenge. *Respiratory Care, 51*(6), 631-636.

Straus, S. E., Glasziou, P., Richardson, W. S., & Haynes, R. B. (2019). *Evidence-based medicine: How to practice and teach EBM.* Elsevier.

Street, R. L., Jr, Makoul, G., Arora, N. K., & Epstein, R. M. (2009). How does communication heal? Pathways linking clinician–patient communication to health outcomes. *Patient Education and Counseling, 74*(3), 295–301.

Stufflebeam, D. L., & Shinkfield, A. J. (2007). *Evaluation theory, models, and applications.* John Wiley & Sons.

The Joint Commission. (2022). Root Cause Analysis. https://www.jointcommission.org/resources/patient-safety-topics/root-cause-analysis/

Thomas, C. M., & Bertram, E. (2019). Role-playing and standardized patients in respiratory care education. *Respiratory Care, 64*(2), 234–239.

Tobin, M. J. (2016). Principles and Practice of Mechanical Ventilation. McGraw-Hill Education.

Tobin, M. J. (2018). Principles and Practice of Mechanical Ventilation. McGraw-Hill Education.

Travis, W. D., Costabel, U., Hansell, D. M., King Jr, T. E., Lynch, D. A., Nicholson, A. G., Ryerson, C. J., Ryu, J. H., Selman, M., Wells, A. U., & Behr, J. (2013). An official American Thoracic Society/European Respiratory Society statement: Update of the international multidisciplinary classification of the idiopathic interstitial pneumonias. American Journal of Respiratory and Critical Care Medicine, 188(6), 733–748. https://doi.org/10.1164/rccm.201308-1483ST

Tuckett, A. (2004). Part 1: The politics of vulnerability. Becoming Vulnerable: Reflections on Pedagogy, Practice, and the Nature of Teaching. https://doi.org/10.4324/9780203566026

U.S. Bureau of Labor Statistics. (2020). Respiratory Therapists. https://www.bls.gov/ooh/healthcare/respiratory-therapists.htm

U.S. Department of Health & Human Services. (2021). Summary of the HIPAA Privacy Rule. https://www.hhs.gov/hipaa/for-professionals/privacy/index.html

U.S. Food and Drug Administration. (n.d.). https://www.fda.gov/

U.S. Government Publishing Office. (2010). Affordable Care Act. https://www.govinfo.gov/app/details/PLAW-111publ148

Vandenhouten, C. L., Malakar, C. L., Kubsch, S., Block, D. E., & Gallagher-Lepak, S. (2017). Continuing Education Needs of Nurses in a Volatile Healthcare Environment. Journal of Continuing Education in Nursing, 48(3), 135–142. https://doi.org/10.3928/00220124-20170215-07

Virchow, J. C., Crompton, G. K., Dal Negro, R., Pedersen, S., Magnan, A., Seidenberg, J., Barnes, P. J., & Group, T. E. S. (2008). Importance of inhaler devices in the management of airway disease. Respiratory Medicine, 102(1), 10–19. https://doi.org/10.1016/j.rmed.2007.07.031

Vygotsky, L. S. (1978). *Mind in society: The development of higher psychological processes.* Harvard University Press.

Walls, R. M., & Murphy, M. F. (2018). Manual of Emergency Airway Management. Lippincott Williams & Wilkins.

Ward, J.J. (2019). High-Flow Oxygen Administration by Nasal Cannula for Adult and Perinatal Patients. Respiratory Care, 64(6), 735-748. https://doi.org/10.4187/respcare.06869

West, J. B. (2016). Respiratory Physiology: The Essentials. Lippincott Williams & Wilkins.

White, G. C., & Kacmarek, R. M. (2017). Clinical Manifestations and Assessment of Respiratory Disease. Elsevier Health Sciences.

Wilkins, R. L., & Stoller, J. K. (2003). *Egan's fundamentals of respiratory care.* Mosby.

Wilkinson, D. J., & Truog, R. D. (2013). *The luck of the draw: physician-related variability in end-of-life decision-making in intensive care.* Intensive Care Medicine, 39(6), 1128-1132.

World Health Organization. (2022). Infection prevention and control. https://www.who.int/infection-prevention/en/

World Health Organization. (2022). Infection prevention and control. https://www.who.int/infection-prevention/en/

Zigmont, J. J., Kappus, L. J., & Sudikoff, S. N. (2011). The 3D model of debriefing: Defusing, discovering, and deepening. Seminars in Perinatology, 35(2), 52–58. https://doi.org/10.1053/j.semperi.2011.01.012

Made in the USA
Monee, IL
07 November 2026

67974631R00050